Springer Theses

Recognizing Outstanding Ph.D. Research

Aims and Scope

The series "Springer Theses" brings together a selection of the very best Ph.D. theses from around the world and across the physical sciences. Nominated and endorsed by two recognized specialists, each published volume has been selected for its scientific excellence and the high impact of its contents for the pertinent field of research. For greater accessibility to non-specialists, the published versions include an extended introduction, as well as a foreword by the student's supervisor explaining the special relevance of the work for the field. As a whole, the series will provide a valuable resource both for newcomers to the research fields described, and for other scientists seeking detailed background information on special questions. Finally, it provides an accredited documentation of the valuable contributions made by today's younger generation of scientists.

Theses may be nominated for publication in this series by heads of department at internationally leading universities or institutes and should fulfill all of the following criteria

- They must be written in good English.
- The topic should fall within the confines of Chemistry, Physics, Earth Sciences, Engineering and related interdisciplinary fields such as Materials, Nanoscience, Chemical Engineering, Complex Systems and Biophysics.
- The work reported in the thesis must represent a significant scientific advance.
- If the thesis includes previously published material, permission to reproduce this must be gained from the respective copyright holder (a maximum 30% of the thesis should be a verbatim reproduction from the author's previous publications).
- They must have been examined and passed during the 12 months prior to nomination.
- Each thesis should include a foreword by the supervisor outlining the significance of its content.
- The theses should have a clearly defined structure including an introduction accessible to new PhD students and scientists not expert in the relevant field.

Indexed by zbMATH.

Daisei Abe

Formation and Evolution of Star-Forming Filaments in Molecular Clouds

Doctoral Thesis accepted by
Nagoya University, Nagoya, Japan

Springer

Author
Dr. Daisei Abe
Nagoya University
Nagoya, Aichi, Japan

Supervisors
Prof. Tsuyoshi Inoue
Department of Physics, Faculty of Science and Engineering
Konan University
Kobe, Aichi, Japan

Prof. Shu-ichiro Inutsuka
Department of Physics, Graduate School of Science
Nagoya University
Nagoya, Aichi, Japan

ISSN 2190-5053 ISSN 2190-5061 (electronic)
Springer Theses
ISBN 978-981-97-6413-6 ISBN 978-981-97-6414-3 (eBook)
https://doi.org/10.1007/978-981-97-6414-3

This Springer imprint is published by the registered company Springer Nature Singapore Pte Ltd.
The registered company address is: 152 Beach Road, #21-01/04 Gateway East, Singapore 189721, Singapore

Supervisors' Foreword

Recent progress in high spatial resolution observations of molecular clouds in the interstellar medium has shown that stars are formed exclusively in dense filamentary regions of molecular clouds. The observed molecular filaments appear to start creating stars once the line-mass of the filament exceeds the critical line-mass (~17 Solar Mass per parsec), at which the thermal pressure in filament cannot support its structure against gravity. This simple and convincing observational evidence directly indicates that we can understand the initial condition of star formation if we know how the filament is created and grows. However, on the theoretical side, many scenarios for the formation of filamentary molecular clouds were proposed by performing various numerical hydrodynamics simulations with different initial conditions and boundary conditions. For example, self-gravity plays the central role in some scenarios while interstellar turbulence was strongly emphasized in others. The existence of significantly different scenario of filament formation has been confusing researchers of star formation processes. Obviously, we should determine the main mechanism for creating star-forming molecular filament.

In 2018, Daisei Abe joined Theoretical Astrophysics Laboratory (TA-Lab) at Department of Physics in Nagoya University, and started his theoretical study of the molecular filament formation using magnetohydrodynamics (MHD) numerical simulations. He focused on the fact that the dense star forming filaments are all formed in shock compressed sheets in previous studies, and started many simulations that reproduced a shock compressed sheet under various setups. He found that the velocity of the shock wave is a crucial parameter that categorizes the resulting filament formation scenarios. Based on this finding, now we can understand why previous studies, done under different shock velocities and different inputs of turbulence, came up with the different scenarios for filament formation. He also extended this study and found the duration of the shock compression is another key parameter that controls star cluster formation.

In the final part of his thesis, he devoted his efforts to analyzing how filaments evolve into massive filaments that may create massive stars. Many groups of observational astronomers suggest that star-forming filaments have a typical width of 0.1 parsec irrespective of their mass per unit length, while numerical simulations generally fail to reproduce it. The observed "universal width of filaments" was challenged by a group that re-analyzed the observations, but its significance seems to remain in most of the observed regions in solar neighborhood. Since the width gives the fragmentation scale of the filament that determines the mass of new-born stars, the physics to determine the filament width must be understood. He has tried to solve this problem in a very unique way by investigating an instability of the MHD slow shock front that bound the filaments. He found that the instability is very active in realistic molecular environments whose scale is characterized by that of ambipolar diffusion (plasma drift). The instability automatically drives active turbulent motion inside the massive filament that provide additional force to keep the filament width against self-gravitational collapse that shrinks the filament width in case without non-thermal support. We believe this thesis provides an important step in understanding the present-day star formation.

We congratulate Dr. Daisei Abe on this outstanding achievement, and we extend our best wishes for further success in his career as a scientist.

Kobe, Japan
Nagoya, Japan
June 2024

Prof. Tsuyoshi Inoue
Prof. Shu-ichiro Inutsuka

Abstract

Recent observations of molecular clouds show that dense filaments are the sites of present-day star formation. Thus, it is necessary to understand the filament formation process because these filaments provide the initial condition for star formation. Theoretical research suggests that shock waves in molecular clouds trigger filament formation. Since several different mechanisms have been proposed for filament formation, the formation mechanism of the observed star-forming filaments requires clarification. In this study, I perform a series of isothermal magnetohydrodynamics (MHD) simulations of filament formation and identify mechanisms for filament formation. I find that the dominant filament formation mode changes with the velocity of the shock wave triggering the filament formation. Moreover, I show that strong shock waves can naturally create high-line-mass filaments such as those observed in massive star-forming regions in a short time.

Observations show that a massive star cluster formation occurs where the peak of gas column density in a cloud exceeds 10^{23} cm^{-2}. I investigate how the initial conditions of massive star formation are realized by performing MHD simulations with gas inflow duration from the boundaries as a controlling parameter. Filaments expand after the duration time for short-duration models, whereas long-duration models lead to star formation by forming massive supercritical filaments. Moreover, when the shock duration is longer than two postshock free-fall times, the peak column density of the compressed layer exceeds 10^{23} cm^{-2}, and the gravitational collapse of the layer leads to the number of OB stars expected to be formed reaches the order of ten (i.e., massive cluster formation).

The filament width is an important quantity because it determines the fragmentation scale by self-gravity. Observations show that the width takes the universal value of 0.1 pc. However, theoretically, the width of the supercritical filaments should contract by self-gravity. Recent studies suggest that massive filaments are bound by the slow-shocks that are caused by accretion flows onto the filaments. As the wavefront of such slow-shock is known to be unstable (slow-shock instability: SSI), the accretion ram pressure can be expected to convert into thermal/turbulent pressure across the shock front that potentially maintains the width. In the scale of dense filaments, ambipolar diffusion (AD) suppresses the SSI at small scales. I investigate the influence of AD on SSI using two-dimensional (2D) MHD simulations. The results demonstrate that the most unstable scale of SSI is approximately five times the length scale of AD calculated using post-shock variables.

Acknowledgments

First of all, I would like to thank my supervisors, Prof. Tsuyoshi Inoue and Prof. Shuichiro Inutsuka. Professor Inoue was my official supervisor from the master's course to the first year of the doctoral program, where he guided my research. Even after he moved as a professor at Konan University, we have had many discussions through online meetings and my business trips to Konan University. I believe that I could complete my doctoral thesis under his guidance. He treated us as one researcher, I could set higher goals for my research. Professor Inutsuka has been my official supervisor since the second year of the doctoral course, but he has had many discussions with me since the master's course. He encouraged us in discussions, so we could reconfirm the importance and interest of our research. I enjoyed my research thanks to his guidance and encouragement.

I would also like to thank my co-authors of the thesis, Prof. Tomoaki Matsumoto, Dr. Rei Enokiya, and Prof. Yasuo Fukui. I learned a great deal from discussions, comments, and revisions in the process of writing the paper. It is thanks to all the co-authors that I could grow as a researcher. The SFUMATO code developed by Prof. Matsumoto was used to produce the important results of this research. I would also like to thank the Japan Student Services Organization. I could spend my research life without financial hardship. The numerical computations were carried out on the XC50 system at the Center for Computational Astrophysics (CfCA) of the National Astronomical Observatory of Japan. This work is supported by Grants-in-aid from the Ministry of Education, Culture, Sports, Science, and Technology (MEXT) of Japan (JP22J15861).

I would like to thank Associate Prof. Hiroshi Kobayashi, Dr. Jiro Shimoda, Mr. Kensuke Kakiuchi, and all the other students who were members of my laboratory at Nagoya University. The guidance and discussions by the lab members had a great influence on shaping my image as a researcher. I would like to thank my colleague Ryunosuke Maeda, with whom I spent my doctoral course, and my junior students Kanta Kitajima and Ryushi Miyayama, who not only discussed our research but also took me out for drinks on our days off. It was thanks to them that I could engage in friendly competition and work hard on my research. I would also like to thank our secretaries, Kyoko Yamazaki, Natsu Kato, Ayako Imai, Ritsuko Watanabe, and Yasuko Iwata for their support.

I would also like to thank Kazunari Iwasaki, Kengo Tomida, Masato I. N. Kobayashi, and Kei K. E. I. Tanaka, and Yoshito Shimajiri, with whom I have had many discussions. I would especially like to thank Kazunari Iwasaki, who taught me a lot of things because our research themes are close, and Kengo Tomida, who is one of the developers of the Athena++ code used in this study, with whom I often consulted on simulation issues. Yuri I. Fujii, Kenji Kurosaki, Shinsuke Takasaka, Doris Arzoumanian, Gabriel Rigon, Keisuke Sugiura, Ryosuke Tominaga, Hidekazu Isoya, Kodai Ito, Yuto Inayoshi, Hana Nishikawa, Yoshiaki Misugi, Yuta Nakanishi, Naoki Nishida, Hiroki Nakatsugawa, and I would like to thank all of my juniors who graduated in that master's course. It was thanks to them that I enjoyed our research discussions and laboratory life. I would also like to thank my friends Katsuka T. Abe, Suchetha Cooray, Hiroto Kondo, Hayato Fukunaga, Kiichi Yoshida, Sayaka Nagasaki, and Shuntaro Yoshida from other theory groups at Nagoya University, with whom we took classes together, competed with each other in master's thesis defense, and had fun daily life.

Professor Ryo Yamazaki, Assistant Professor Yutaka Ohira, and Shota Kisaka, who were my mentors at the Yamazaki Laboratory of Aoyama Gakuin University when I was an undergraduate student, recommended me to enter Nagoya University and supported my decision when I decided to study star formation. I learned a lot from their severe comments on my research at that time. In particular, Ryo Yamazaki was my first supervisor and the first researcher with whom I had discussions. His attitude of enjoying research influenced me, and it was then that I decided to become a researcher. I would like to thank my undergraduate classmates Takafumi Kobayashi, Natsuki Ishizaka, Shoma Kamijima, Masashi Komai, Shuto Sei, Shun Masumoto, Yuto Obara, and Shin Kakuchi. I was anxious about being in the laboratory environment for the first time, but thanks to them I could enjoy my research and life in the laboratory. I would like to thank my friends Keisuke Yamamoto, Yukimaru Saito, Saki Ikeda, and all my friends from high school for their encouragement.

Finally, I would like to express my gratitude to my family. My father, Kazuki, is a bar owner, and to improve the service at his bar, he became a wine and sake sommelier. I sometimes helped him at his bar before he became a sommelier, and he found time to study for his certification while working. He is a person to look up to because of his strong commitment to what he decides to do. I was encouraged by his attitude toward work. My mother, Yukari, supported me more than anyone else. When I told her that I wanted to become a researcher, she was a little apprehensive, but she believed in me and let me do what I wanted to do. My younger brother, Issa is studying to become a fashion designer. Becoming a fashion designer seems to be difficult, and he is taking on the challenge. I am encouraged by the way he is growing up. I would like to thank my grandparents and relatives. They have always watched my growth and supported me.

Contents

Chapter 1
Introduction

Star formation and evolution is a fundamental physical process in molecular clouds in galaxies. Furthermore, stars cause stellar winds and supernovae, which affect galaxy evolution and supply the gas that is the ingredient of stars. Since the evolution of a star is determined by its mass, the origin of the initial mass function is important, which implies that it is important to understand the star formation process. Therefore, understanding the star formation process is essential to understanding the universe as a whole. However, the initial conditions of the star formation process have not yet been unknown. In this thesis, we investigate the formation and evolution of dense filamentary molecular clouds, sites of star formation, to understand the initial condition of star formation.

1.1 Overview of Molecular Cloud

1.1.1 Molecular Clouds Formation

In galaxies, there is multi-phase gas such as hot ionized medium, warm/cold neutral medium, molecular clouds, and HII region. We focus on warm/cold neutral medium because the former transit to molecular clouds and the latter are sites of star formation. The warm neutral medium is neutral hydrogen atomic gas whose density and temperature of $\sim$0.1 cm^{-3} and $\sim$8000 K, respectively. Almost all of the volume of a galaxy is occupied by the warm neutral medium and the hot ionized medium. The cold neutral medium is also neutral hydrogen atomic gas whose density and temperature of $\sim$1 cm^{-3} and $\sim$100 K, respectively. These two components can coexist because pressure equilibrium with $\sim$1000 K cm^{-3} is realized between the warm and cold neutral medium. Therefore the warm/cold neutral medium is stable for linear perturbation. The molecular clouds are mainly composed of molecular hydrogen and range in size from a few pc to 100 pc. The density and temperature are more than

D. Abe, *Formation and Evolution of Star-Forming Filaments in Molecular Clouds*,
Springer Theses, https://doi.org/10.1007/978-981-97-6414-3_1

$10^2\,\text{cm}^{-3}$ and $\sim 10\,\text{K}\,\text{cm}^{-3}$, respectively, that is pressure is larger than the warm/cold medium due to self-gravity.

We introduce how dense gas formation such as the cold neutral medium and the molecular clouds from diffuse atomic gas is achieved. The cooling and heating processes are essential to understand dense gas formation. Koyama and Inutsuka (2000) shows an important cooling and heating process in interstellar medium. The heating mechanisms are the photoelectric emission from dust grains and PAHs (Polycyclic Aromatic Hydrocarbons), ionization by cosmic rays and soft X-rays, and the formation and photo-dissociation of H_2. The local far-ultraviolet (FUV) field is set to 1.7 times Habing's estimate (Wolfire et al. 1995). The dominant cooling processes are the line emission from H, C, O, Si, and Fe, by rotation/vibrational line from CO, as well as by atomic and molecular collisions with dust grains.

Adiabatic fluid is stable for an increment of density because of pressure enhancement. However, if there is a cooling process, it can be unstable. Field (1965) studies the stability conditions of a uniform medium about heating and cooling and shows that a phase transition (warm neutral medium becomes cold one) due to thermal instability occurs when the following conditions are satisfied

$$\left(\frac{\partial \mathcal{L}}{\partial T}\right)_{\text{P}} < 0 \Leftrightarrow \left(\frac{\partial P}{\partial \rho}\right)_{\mathcal{L}} < 0 \tag{1.1}$$

where, $\mathcal{L} = \rho\Lambda - \Gamma$, $\rho\Lambda$ denote the cooling function per volume and Γ denotes the heating function. This instability occurs for temperatures between $\sim$100 and $\sim$5000 K in the interstellar medium. Thermal conduction stabilizes the perturbations with smaller wavelengths. The critical wavelength is

$$\lambda_{\text{c}} = 2\pi \left\{ \frac{\rho}{K} \left[\frac{\partial \mathcal{L}}{\partial T} \right]_{\text{P}} \right\}^{-1/2} \sim \sqrt{\frac{KT}{\rho^2 \Lambda}} \tag{1.2}$$

where K denotes the coefficient of thermal conduction. Cold neutral medium around this scale forms. The molecular cloud that eventually forms by this instability has a clumpy structure.

As we mentioned, the warm/cold neutral medium is stable for linear perturbation. Thus, a nonlinear trigger is necessary to create the cold neutral medium. One of the triggers is a shock wave. A shock wave is caused by gas collision, feedback from massive stars including supernovae and expanding HII regions, and encounters with galactic spiral shock. Inoue and Inutsuka (2009) indicates that the direct formation of molecular clouds from a warm neutral medium is not a typical molecular cloud formation process unless the direction of shock wave propagation is biased to the orientation of the mean magnetic field. Thus, they emphasize the importance of multiple compression, as a typical formation process of molecular clouds.

1.1.2 Properties of Molecular Clouds and Star Formation

1.1.2.1 Isothermal Approximation

In molecular clouds, the dominant heating source is cosmic rays. Dominant cooling sources are CO rotation/vibration line emission and dust cooling at higher densities ($> 10^5\ \mathrm{cm}^{-3}$). The balance between them maintains a temperature of approximately 10 K (sound speed $c_s = 0.2\ \mathrm{km\ s}^{-1}$). Using the cooling rate at the thermal equilibrium of the molecular gas, the cooling time t_{cool} can be written as

$$t_{\mathrm{cool}} = \frac{k_{\mathrm{B}}T}{n\Lambda} \sim 10^4\ \mathrm{yr}, \tag{1.3}$$

for $n = 100\ \mathrm{cm}^{-3}$ where k_{B} is Boltzmann constant. This timescale is smaller than the dynamical timescale of molecular clouds. Therefore, the isothermal equation of state is justified in the molecular cloud.

1.1.2.2 MHD Approximation

The main component of molecular clouds is molecular hydrogen, i.e., uncharged particles, and there is a slightly ionized component (ionization degree of $\sim 10^{-7}$). However, these neutrals also interact indirectly with the magnetic field like charged particles, and as a result, the motion of the gas in the molecular cloud can be approximated to follow the ideal magneto-hydrodynamics (MHD) equation. The qualitative explanation is as follows. Electrons and protons have electric charges, and in the presence of a magnetic field, they are subject to the Lorentz force and orbit around the magnetic field lines. From a macroscopic point of view, a group of charged particles (plasma) couples with a magnetic field (magnetic frozen-in). On the other hand, neutral particles such as molecular hydrogen interact with charged particles via collisions. When plasma moves, the group of neutrals (neutral fluid) is dragged via collision with plasma. Since plasma couples with the magnetic field, The motion of neutral fluid, the magnetic field, and plasma couple. Therefore, molecular clouds behave just like a fully ionized plasma, and their motion can be described by the MHD equation.

We consider the specifics using the equations. The Lorentz force per unit volume is

$$\mathrm{f}_{\mathrm{L}} = \frac{1}{4\pi}(\nabla \times \mathbf{B}) \times \mathbf{B}, \tag{1.4}$$

where, $\mathbf{B}$ denotes magnetic field. The frictional force by ion fluid[1] on neutral fluid can be written as follows

$$\mathbf{f}_{\mathrm{d}} = \gamma_{\mathrm{in}}\rho_{\mathrm{n}}\rho_{\mathrm{i}}\left(\mathbf{u}_{\mathrm{i}} - \mathbf{u}_{\mathrm{n}}\right) \tag{1.5}$$

where, ρ_{n} and ρ_{i} denote the mass density for the neutral fluid and the ion fluid, respectively. When collision velocity between an ion and a neutral molecule is small, i.e., cold environment such as a molecular cloud, Osterbrock (1961) pointed out that Langevin cross section is correct for the cross-section between ions and neutrals, because the ion induces a dipole moment in the neutral molecule, effectively changing the collision cross-section. For the Langevin cross-section, since $\sigma_{\mathrm{in}} \propto w^{-1}$, γ_{in} becomes constant. Draine et al. (1983) computed the resulting value for $\gamma_{\mathrm{in}} = 3.5 \times 10^{13}\mathrm{cm}^3\mathrm{g}^{-1}\mathrm{s}^{-1}$. The interior of a molecular cloud is partially ionized by cosmic rays. If the ionization rate for a neutral particle is ζ, the ionization rate of the neutral particle per unit volume is ζn_{n}. On the other hand, the recombination rate of ions and electrons is proportional to the number density of both ions and electrons and is, therefore, proportional to $n_{\mathrm{i}}n_{\mathrm{e}} \propto n_{\mathrm{i}}^2$. Considering the steady state, the ionization and recombination are balanced and can be written using the mass density and the constant $C = 3 \times 10^{-16}\ \mathrm{cm}^{-3/2}\ \mathrm{g}^{1/2}$ as follows

$$\rho_{\mathrm{i}} = C\rho_{\mathrm{n}}^{1/2}. \tag{1.6}$$

Since the gas in the molecular cloud is dense enough to collide between gas particles, the Lorentz force and the frictional force are dominant over the other forces (gravity and pressure gradient forces) and their sum is nearly zero. Therefore, considering $\mathbf{f}_{\mathrm{L}} = \mathbf{f}_{\mathrm{d}}$, the ion-neutral drift velocity is

$$\mathbf{v}_{\mathrm{d}} \equiv \mathbf{u}_{\mathrm{i}} - \mathbf{u}_{\mathrm{n}} = \frac{1}{4\pi\gamma\rho_{\mathrm{n}}\rho_{\mathrm{i}}}(\nabla \times \mathbf{B}) \times \mathbf{B} \tag{1.7}$$

To evaluate the drift velocity, we estimate the order of Eq. 1.7,

$$v_{\mathrm{d}} \sim \frac{B^2}{4\pi\gamma\rho_{\mathrm{n}}\rho_{\mathrm{i}}L} \approx \frac{v_{\mathrm{A}}^2}{V}, \tag{1.8}$$

where, $V \equiv \gamma\rho_{\mathrm{i}}L = \gamma C L\rho^{1/2}$. $v_{\mathrm{A}} \equiv B/\sqrt{4\pi\rho}$ denotes Alfvén velocity. We assume $\rho = \rho_{\mathrm{n}} + \rho_{\mathrm{i}} \approx \rho_{\mathrm{n}}$. When $L = 1$ pc, $n_{\mathrm{n}} = 100\ \mathrm{cm}^{-3}$, and $B = 10\ \mu\mathrm{G}$, then $V = 19\ \mathrm{km\,s^{-1}}$ and $v_{\mathrm{A}} = 1.4\ \mathrm{km\,s^{-1}}$. Therefore, $v_{\mathrm{d}} \sim 0.1\ \mathrm{km\,s^{-1}} \ll v_{\mathrm{A}}$ is satisfied, indicating that the difference in motion between plasma and neutral fluid is small and the MHD approximation is justified. Note that we cannot neglect the effect of the slip of the magnetic field at the small scale.

[1] Ions, which are the most massive of the charged particles, are responsible for momentum exchange in collisions with neutral particles. Electrons have less mass than ions, so they contribute little to the momentum exchange. Electrons do not contribute much to momentum exchange because they have less mass than ions.

1.1.2.3 Supersonic Turbulence

One of the important features given by observation of ^{12}CO J $= 1 - 0$ is the line width which is larger than the expected line width by the sound speed of gas, indicating that molecular clouds are super-sonically turbulent. There is typical relation of $\delta v \propto l^{0.5}$ (Larson 1981). The super-sonic turbulence creates density inhomogeneity with a power spectrum of $(\log \rho)_k^2 \propto k^{-4}$ (Beresnyak et al. 2005; Elmegreen and Scalo 2004; Scalo and Elmegreen 2004).

1.1.2.4 Shock Wave

We should emphasize that shock waves are important for the dynamics of molecular clouds. As mentioned in Sect. 1.1.1, propagation of shock waves is ubiquitous in galaxies. Using supernovae as an example, we can estimate the time scale of shock wave sweeping for whole gas in a galactic disk,

$$t_{\rm swept} \sim \frac{V_{\rm galaxy}^{1/3}}{v_{\rm SN} N_{\rm SN}^{1/3}} \simeq 1 \text{ Myr} \tag{1.9}$$

where, $V_{\rm galaxy} \sim (10 \text{ kpc})^2 \times 100$ pc, $v_{\rm SN} \sim 100$ km s^{-1}, and $N_{\rm SN} \sim 10^6 \text{ yr}/100 \text{ yr}$ denote the volume of galactic disk, typical shock velocity of supernovae, and the number of supernovae of age one million year, respectively. We assume supernova explosions occur once every 100 yr in a galaxy. Therefore, molecular clouds are shock-compressed once per million years.

Inutsuka et al. (2015) found the resultant star formation efficiency $\epsilon_{\rm SF} \sim 10^{-2}$ within a cloud, which is consistent with observation (Zuckerman and Evans 1974), based on the episodes of multiple supersonic compression. Zucker et al. (2022) reported an analysis of the 3D positions, shapes, and motions of dense gas and young stars within 200 pc of the Sun from *Gaia* data. They found that the most star-forming regions lie on the surface of the shock front of multiple supernovae (Local Bubble) and that their young stars show outward expansion mainly perpendicular to the bubble's surface. These findings support the scenario that the expansion of the Local Bubble swept up the interstellar medium and drives star formation, in turn providing robust observational support for the theory of supernova-driven star formation (Inutsuka et al. 2015). When molecular clouds are shock-compressed, further dense structure formation occurs. This topic is our main target, details are shown in Sect. 1.2.

1.1.2.5 Self-Gravitational Collapse in Molecular Clouds

Within the high-density region of a molecular cloud, a particularly dense region that is the parent body of a star is called a molecular cloud core. When the molecular cloud core gravitationally contracts, it evolves into a denser object, and eventually, hydrogen-burning reactions begin. The central body is then pressurized by the nuclear reaction energy, the gas sphere stops contracting and the radius of the central body remains constant. The central body at this stage is in the main sequence and can be said to have given birth to a star.

1.2 Filamentaly Molecular Clouds

Stars are formed in dense regions in molecular clouds (e.g., Lada et al. 2010; Enoch et al. 2007; André et al. 2014; Hennebelle and Inutsuka 2019). Recent dust thermal emission observations by the *Herschel* space telescope (band: 70–500 μm) revealed that dense filamentary structures are ubiquitous in nearby molecular clouds (e.g., André et al. 2010; Arzoumanian et al. 2011). Additionally, star-forming cores and young stellar objects are embedded along the filaments, which indicates their crucial role in star formation (Könyves et al. 2015). Therefore, comprehension of the formation and evolution mechanisms is significant to understand the star formation scenario and the initial condition of star formation.

1.2.1 *Definitions*

To understand the stability of filament, let us consider a uniform, infinitely elongated, axisymmetric, and cylindrical gas. The Poisson equation for self-gravity is

$$\nabla^2\phi = \frac{1}{r}\frac{d}{dr}r\frac{d\phi}{dr} = 4\pi G\rho, \tag{1.10}$$

where, ϕ, G is gravitational potential and constant, respectively. To get the gravitational force, multiplying r and integrate from $r = 0$ to outer edge of cylinder $r = R$,

$$R\left.\frac{d\phi}{dr}\right|_{r=R} = 2G\int_0^R 2\pi r\rho dr = 2GM_{\mathrm{line}}, \tag{1.11}$$

and,

$$F_{\mathrm{g}} = \left.\frac{d\phi}{dr}\right|_{r=R} = \frac{2GM_{\mathrm{line}}}{R} =\propto R^{-1}, \tag{1.12}$$

where, $M_{\mathrm{line}} \equiv \int_0^R 2\pi r\rho dr$ is mass per unit length (Line-mass). On the other hand, pressure gradient force is

$$F_{\mathrm{p}} = \rho^{-1}\left.\frac{\partial p}{\partial r}\right|_{r=R} \propto R^{-2\gamma_{\mathrm{eff}}+1}, \tag{1.13}$$

where, $p = K\rho^{\gamma_{\rm eff}}$ is pressure, $\gamma_{\rm eff}$ is effective specific ratio. The ratio of Eqs. 1.12 and 1.13 is

$$\frac{F_{\rm p}}{F_{\rm g}} = R^{2(1-\gamma_{\rm eff})}. \tag{1.14}$$

For the isothermal cylindrical system ($\gamma_{\rm eff} = 1$), it is possible to become dynamical equilibrium when a filament has the critical line-mass. The hydro-static equilibrium distribution is given by

$$\rho_{\rm eq}(r) = \rho_c \left[1 + \left(\frac{r}{H_0}\right)^2\right]^{-2}, \tag{1.15}$$

where, $\rho_{\rm c}$ and $H_0 \equiv \sqrt{\frac{2c_{\rm s}^2}{\pi G\rho_{\rm c}}}$ are central density and scale-height. To create stars in filaments, their line-masses have to exceed the critical line-mass $M_{\rm line,cr}$ for gravitational instability,

$$M_{\rm line,cr,th} = \int_0^\infty 2\pi\rho_{\rm eq}(r) r dr = 2c_{\rm s}^2/G \simeq 17\ {\rm M_\odot\ pc^{-1}}, \tag{1.16}$$

where, $c_{\rm s} \simeq 0.2$ km s^{-1} is the isothermal sound speed of typical molecular clouds (e.g., Stodólkiewicz 1963; Ostriker 1964; Inutsuka and Miyama 1992a; Inutsuka and Miyama 1997). Non-thermal velocity dispersion Δv, i.e., turbulence, makes the critical line-mass increase. Equation 1.16 can be written as

$$M_{\rm line,cr,tot} = 2\left(c_{\rm s}^2 + \Delta v^2\right)/G \simeq 17 + 425 \left(\frac{\Delta v}{1\ {\rm km\ s^{-1}}}\right)^2\ {\rm M_\odot\ pc^{-1}}. \tag{1.17}$$

The magnetic field also makes the critical line-mass amplify. It is known that star-forming filaments are threaded by magnetic field perpendicularly (see details in Sect. 1.2.4.3). Tomisaka (2014) gave the formula of line-mass including the effect of the magnetic field is

$$M_{\rm line,cr,B} \simeq 2.2 \left(\frac{w}{0.1{\rm pc}}\right)\left(\frac{B}{10\mu{\rm G}}\right) M_\odot{\rm pc}^{-1} + 13.9 \left(\frac{c_s}{0.19\ {\rm km\ s^{-1}}}\right)^2 M_\odot{\rm pc}^{-1}, \tag{1.18}$$

where, w and B are the widths of a filament and magnetic field strength.

The column density profiles are parameterized by Plummer function,

$$N(x) = A_{\rm p} \frac{n_0 R_{\rm flat}}{\left(1 + (x/R_{\rm flat})^2\right)^{(p-1)/2}}, \tag{1.19}$$

where, $A_{\rm p}$ and x are constant and projected distance from filament's axis. $R_{\rm flat} \equiv N(0)/(n_0 A_{\rm p})$. The equilibrium filament width i.e., full-width-half-maximum of the

Plummer function is given by

$$w_{\mathrm{eq}} = 2R_{\mathrm{flat}}\left(2^{2/(p-1)} - 1\right)^{1/2}. \tag{1.20}$$

For $p = 4$, Eq. 1.19 corresponds isothermal case, corresponding 3D profile is Eq. 1.15, $A_{\mathrm{p}} = \pi/2$, $R_{\mathrm{flat}} = H_0$, and $w_{\mathrm{eq}} \simeq 1.53R_{\mathrm{flat}}$.

1.2.2 Formation

Previous theoretical research on filament formation proposed several types of mechanisms. We summarize the filament formation mechanism in the following, Fig. 1.1, Table 1.1.

1.2.2.1 Type G

The first one is a well-known self-gravitational fragmentation of a sheetlike cloud (Tomisaka and Ikeuchi 1983; Miyama et al. 1987a, b; Nagai et al. 1998; Kitsionas and Whitworth 2007; Balfour et al. 2015, 2017), which is created when molecular clouds are shock-compressed. In this paper, we call this type of filament formation type G formation. According to linear stability analysis conducted by Nagai et al.

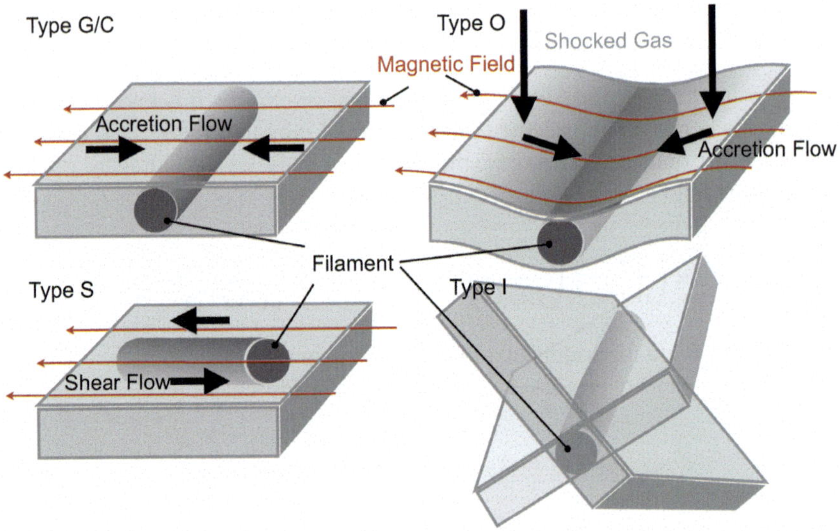

Fig. 1.1 Schematic illustrations of filament formation mechanisms (Abe et al. 2021; Pineda et al. 2022). See also Table 1.1

Table 1.1 Filament formation mechanisms

Category	Filament versus magnetic field	A brief description of the formation mechanism
Type G	Perpendicular	Self-gravitational fragmentation of a sheetlike cloud
Type I	–	The intersection of two shocked layers
Type O	Perpendicular	The effect of an oblique MHD shock wave
Type C	Perpendicular	Converging gas flow components along the magnetic field in the post-shock layer
Type S	Parallel	Stretching by turbulent shear flows, small line-mass

(1998), gravitational instability creates filaments with line-masses larger than but comparable to the critical line-mass when the width of the sheet is comparable to the Jeans length (i.e., the self-gravity is important in the dynamics).

1.2.2.2 Type I

Padoan and Nordlund (1999), Pudritz and Kevlahan (2013), Matsumoto et al. (2015), and, Federrath (2016) showed filament formation induced by turbulence in molecular clouds by using three-dimensional magnetohydrodynamic (MHD) simulation. Padoan and Nordlund (1999) reported that the turbulent velocity given in an initially uniform molecular gas induces the formation of shock-compressed sheets, and then the interaction of two sheets creates a filament at the intersection of them. We call this mechanism type I filament formation. It should be noted that the type I process occurs in the super-Alfvénic case, in which the initial Alfvénic Mach number $\mathcal{M}_{\mathrm{A,i}} \gtrsim 10$.

1.2.2.3 Type O

Inoue and Fukui (2013), Vaidya et al. (2013), Inutsuka et al. (2015), and Inoue et al. (2018) reported that the filaments are generated when a shock wave sweeps a cloud containing density inhomogeneity or clumps. In this case, a dense blob embedded in a magnetized molecular cloud is transformed into a dense filament in the shock-compressed layer owing to the effect of an oblique (or curved) MHD shock wave. Throughout this study, this filament formation mode is called type O mode. It should be noted that the effects of thermal instability and turbulence cause molecular clouds to be highly clumpy by nature (e.g., Inoue and Inutsuka 2012a). Note that in most theoretical works that study structure formation in molecular clouds by turbulence, a uniform-density gas is used as an initial condition. Therefore the type O mode is missing in the studies that assume the uniform initial density. As we show subsequently, the type O mode dominates over other modes when the shock Mach number is high.

1.2.2.4 Type C

Chen and Ostriker (2014) and Chen (2015) showed another filament formation behind shock waves that is different from type O. They found that filaments are formed by converging gas flow components along the local magnetic field in the case the substantial turbulent motions are given in the initial condition. We call this mode type C filament creation mechanism. This mechanism creates the filaments that are perpendicular to the background magnetic field when the turbulence is sub-Alfvénic (Chen and Ostriker 2014; Chen 2015; Planck Collaboration XXXV 2016). This is because the component of turbulent velocity perpendicular to the magnetic field is suppressed by magnetic tension force, and only motion parallel to the magnetic field can compress the gas. Padoan and Nordlund (1999) also reported the type C process in the trans-Alfvénic case ($\mathcal{M}_{\rm A,i} \sim 1$).

1.2.2.5 Type S

Hennebelle (2013) reported that a small clump in a turbulent molecular cloud is stretched by turbulent shear flows and evolves into a small line-mass filament parallel to the magnetic field (see also, Inoue and Inutsuka 2016, for the origin of HI filament/fiber). We call this mode type S mechanism, and this type S creates filaments that are parallel to the magnetic field lines in contrast to the other types that create filaments perpendicular to the magnetic field lines. Xu et al. (2019) showed that the anisotropic nature of turbulent MHD eddies results in filamentary structure formation parallel to a magnetic field.

1.2.3 Evolution

1.2.3.1 Accretion

Gas accretion onto filaments is a crucial process to determine the evolution and physical properties such as line-mass and width of filaments. Klessen and Hennebelle (2010) argued that turbulence is generated by the accretion in a general sense and that turbulence affects evolution. As mentioned in Sect. 1.2.2, the formation mechanisms of high-density filaments, which are important for star formation, are mainly Type G, C, and O. The common feature of these mechanisms is that filaments are formed by gas flowing perpendicular to the long axis of the filament (along the magnetic field) in a shocked-compressed sheet. Line observations also show the evidence of the perpendicular accretion onto filaments (Palmeirim et al. 2013; Shimajiri et al. 2019; Chen et al. 2020). Especially, Shimajiri et al. (2019) claimed that the accretion onto filaments in the shocked sheet. The accretion rate $\dot{M}_{\rm line}$ is estimated to range from a few 10 to a few 100 $M_\odot$ Myr^{-1} pc^{-1} (Palmeirim et al. 2013; Bonne et al. 2020). The accretion timescale can be written as

$$t_{\rm acc} = \frac{M_{\rm line}}{\dot{M}_{\rm line}}, \tag{1.21}$$

which is the order of ~ 0.1 Myr.

Shock wave with the Alfvén Mach number $\mathcal{M}_{\rm A} < 1$ and sonic Mach number $\mathcal{M}_{\rm s} > 1$ is called "slow (mode) shock." Given that the massive filaments are formed in the post-shock layer threaded by the strong magnetic field, the surface of the filament is naturally bound by the slow-shocks. Lessen and Deshpande (1967) found through linear stability analysis that the slow-shock front is corrugationally unstable (slow-shock instability, hereafter SSI). The mechanism of SSI is as follows. For the slow-shock, in contrast to the fast shock, the component of the magnetic field tangential to the shock surface decreases across the front. Thus, when the shock corrugates, magnetic field lines kink as red lines shown in Fig. 1.2. Because gas flows along the magnetic field, the gas converges behind the peak of the shock, while it diverges behind the valley. Such flow patterns increase (decrease) the pressure behind the peaks (valleys), which further push up (pull down) the shock front unstably. Édel'Man (1989) showed that the approximated dispersion relation of SSI for $\mathcal{M}_{\rm A} \ll 1$ can be written as

$$\omega \sim -i\frac{\gamma - 1}{\gamma + 1} M_{\rm A}^2 v_{\rm sh} k, \tag{1.22}$$

where, ω, $v_{\rm sh}$, and k are frequency, shock velocity, and wave number of the shock corrugation. As an more accurate solution, Édel'Man (1989) derived the approximate dispersion relation for $\gamma = 5/3$:

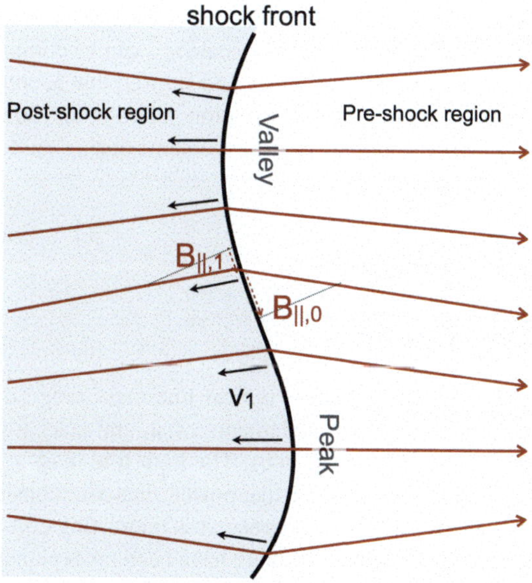

Fig. 1.2 Schematic illustration of two-dimensional slow mode shock. Reproduced from Abe et al. (2024).

$$\omega \simeq -i\left(0.208M_{\rm A}^2 - 0.0775M_{\rm A}^4 + 0.06M_{\rm A}^6\right) v_{\rm sh}k. \quad (1.23)$$

Equations (1.22) and (1.23) show that the most unstable scale is infinitesimally small. This unphysical feature stems from ideal approximation and resulting discontinuous treatment of the shock. To know the physical scale length of the SSI, we have to take a non-ideal effect into account.

1.2.3.2 Fragmentation

Understanding the fragmentation of filaments is important since fragmentation results in core formation. Linear analysis by Stodólkiewicz (1963); Inutsuka and Miyama (1992b) shows the most unstable wavelength for self-gravitational fragmentation of filaments $\lambda_{\rm max}$,

$$\lambda_{\rm max} \sim 4 \times w_{\rm eq}. \quad (1.24)$$

The distance collects gas is typically four times its width when a filament fragments. Thus, (sub-) critical filaments will form equally spaced cores. The e-folding time for most unstable mode ($\lambda = \lambda_{\rm max}$) is

$$t_{\rm frag} = \frac{3}{2\sqrt{\pi G \rho_0}} = 1.66\left(\frac{n_0}{10^3\ {\rm cm}^{-3}}\right)^{-1/2} {\rm Myr}. \quad (1.25)$$

However, the actual spacing of cores given by observations is not consistent with theories. The spacing is irregular or shorter than $\lambda_{\rm max}$ or scattered. This complexity might come from the accretion onto filaments considering the accretion timescale $t_{\rm acc}$ and the fragmentation timescale $t_{\rm frag}$ can be comparable. Clarke et al. (2016) performed numerical simulations and showed that accretion affects the dispersion relation for gravitational fragmentation. Chira et al. (2018) emphasized the importance of turbulence and density enhancements during the formation of filaments.

1.2.4 Properties

1.2.4.1 Line Mass

André et al. (2019) show typical line-mass range of 5–17 $M_\odot$ pc^{-1}. Both super-critical and sub-critical filaments exist, and stars form in at least thermally super-critical ($M_{\rm line}$ $M_{\rm line,cr,th}$) ones. The high line-mass region ($M_{\rm line} \gtrsim 17\ M_\odot\ {\rm pc}^{-1}$) is consistent with a Salpeter-like power, thus also consistent with initial mass function (IMF) and (CMF), in this sense, it is interesting considering the link among FLMF, IMF, and CMF. However, there have been no explanations. In Chap. 2, we give hints to solve the origin of FLMF.

Infrared Dark Clouds (IRDCs) are the sites for massive mass star formation typically located at distances of a few kpc and have also high line-mass ($\gtrsim$ 100 $M_\odot$ pc^{-1}) filamentary structures (e.g., Peretto et al. 2014; Henshaw et al. 2014). Fukui et al. (2019); Tokuda et al. (2019) show that massive filaments of $\gtrsim$ 100 $M_\odot$ pc^{-1} exist in the Magellanic Clouds using ALMA observations.

Arzoumanian et al. (2018); André et al. (2019) revealed that *Herschel* Gould Belt filaments have typically 0.5 pc length. They also show the histogram of the 599 filaments' width. The mean value is 0.1 pc with a spread by a factor of 2. Similar results are given by $C^{18}O$ line emission observations (Orkisz et al. 2019; Suri et al. 2019). The widths given by using a denser tracer tend to be smaller than 0.1 pc, but these observations reflect the internal sub-structure in filaments. It is noted that the dynamic range of the dust emission is wider than the line emission, therefore the dust observations provide the real filament's width. Arzoumanian et al. (2018) also shows the relation between column density (line-mass) and the widths of filaments in nearby clouds. The remarkable point is that the widths of filaments are 0.1 pc regardless of their line-mass, and are inconsistent with the Jeans length. The width of massive filaments should depend on column density (line-mass) due to the strong self-gravity and should be smaller than 0.1 pc. All theoretical models and simulations could have not explained the observational fact (Fischera and Martin 2012; Federrath 2016; Ntormousi and Hennebelle 2019). Fischera and Martin (2012) considered the external pressure and calculated the dynamical equilibrium of a filament. They concluded that sub-critical filaments have a width of 0.1 pc. However, they did not explain the maintaining mechanism of the width of super-critical filaments. This contradiction between the theories and the observational fact is one of the open questions in the star formation domain remaining about 10 years. In this thesis, we call this problem "0.1 pc problem," and discuss it in Chap. 4.

1.2.4.2 Radial Profile

The radial profile of filaments provides information about their stability, equation of state, and environments. As mentioned in Sect. 1.2, the index of $p = 4$ corresponds to the isothermal case. However, most filaments show $p = 1.5$–2.5 (Arzoumanian et al. 2011; Palmeirim et al. 2013), indicating that the isothermal equilibrium cannot explain the feature of the observed filaments. Tomisaka (2014); Kashiwagi and Tomisaka (2021) calculated the equilibrium solution of a filament laterally threaded by the magnetic field and showed the index $2.48 \lesssim p \lesssim 2.73$. Pressure-confined filaments also have flatter profile $p < 4$ (Fischera and Martin 2012).

1.2.4.3 Magnetic Field

Observation of polarized dust emission or stellar polarization indicated that the direction of the magnetic field is perpendicular to the long axis of the high column density filaments (Palmeirim et al. 2013; Planck Collaboration XXXV 2016; Soler and

Hennebelle 2017). For low column density filaments ($N \lesssim 10^{21-22}$ cm^{-2}), on the other hand, the long axis of filaments tends to orient parallel to the magnetic field. These facts can be interpreted considering the mechanism of filament formation (see Sect. 1.2.2).

Davis-Chandrasekhar-Fermi method and the Zeeman observations derive plane-of-sky and line-of-sight magnetic field strength (Davis 1951; Chandrasekhar and Fermi 1953; Crutcher 2012). The typical magnetic field strength is 70 μG for nearby filaments and 45–500 μG for IRDC. Note that the observations of the magnetic field strength have limitations or uncertainty.

1.2.4.4 Turbulence

The gas velocity dispersions in filaments are measured by line observations. Arzoumanian et al. (2013) shows that the velocity dispersion in filaments close to sound speed and finds the relation between total velocity dispersion and line-mass such as $\sigma_{\rm tot} \propto M_{\rm line}^{0.36} \pm 0.07$, where $\sigma_{\rm tot} \equiv \sqrt{c_s^2 + \Delta v^2}$.

Simulations show that turbulence within a filament is driven by gas accretion from the outside of the filament (Seifried and Walch 2015; Clarke et al. 2017; Heigl et al. 2018). However, Heigl et al. (2020) indicated that turbulence does not contribute to the stability of filaments because the profile of turbulent ram pressure is flat. Furthermore, it is known that turbulence kinetic energy results in a only few percent of the kinetic energy of the accretion flow (Klessen and Hennebelle 2010). The resulting velocity dispersion is trans-sonic to supersonic (Seifried and Walch 2015; Clarke et al. 2017; Heigl et al. 2018, 2020), which agrees with observations (Arzoumanian et al. 2013). Therefore, the statement in previous works is that gas accretion has nothing to stabilize the radial collapse of filaments.

1.3 Purpose of This Thesis

This study aims to understand the initial conditions of the star formation process. We investigate the formation and evolution of filamentary molecular clouds. Previous theoretical research on filament formation proposed many types of mechanisms as shown in Sect. 1.2.2. However, it is still unclear which type is responsible for the creation of star-forming filaments. Most of the proposed mechanisms (type I, O, C, and possibly type G) are triggered by shock compression. Thus, in this study, we perform a series of isothermal MHD simulations of filament formation by shock waves, focusing on the influences of shock strength, turbulence, and self-gravity on the filament formation mechanism. Note here that, since type I is effective only in highly super-Alfvénic turbulence, we hardly observe it in the results of our simulations that are performed under a realistic 10 μG initial magnetic field.

Previous studies have demonstrated that strong MHD shock compression allows for the formation of massive cores (Inoue and Fukui 2013; Inoue et al. 2018; Sakre et al. 2021). However, these studies implicitly assumed a long duration of a shock wave (corresponding to collisions between large clouds), whereas the realistic duration of a shock wave depends on the situation, e.g., the shock created by cloud collision cannot be kept over a crossing time. On the other hand, simulations of unmagnetized cloud collisions by Takahira et al. (2014) demonstrated that gravitationally bound cores do not form in the case of short duration. Therefore, a systematic study into the effect of shock duration on the resulting filament/core formation is required. Furthermore, Enokiya et al. (2021) demonstrated that the peak column density of a star-forming region correlates with the number of OB stars in the system, and massive star clusters with more than 10 OB-type stars are associated with massive clouds whose peak column density exceeds 10^{23} cm^{-2}. Thus, the physical origin of the threshold peak column density of 10^{23} cm^{-2} must be clarified. In Chap. 3, we show the effect of shock wave duration on star formation.

The width of filaments is one of the significant quantities to determine the initial condition for star formation. In linear theory, the self-gravitational fragmentation length scale of the filament depends on the width (Stodólkiewicz 1963; Inutsuka and Miyama 1992a). Arzoumanian et al. (2018) revealed that *Herschel* Gould Belt filaments have characteristic width of 0.1 pc. Similar results are given by $C^{18}O$ line emission observations (Orkisz et al. 2019; Suri et al. 2019). Remarkably, the filaments keep their width regardless of their line-mass even beyond 100 $M_{\odot}$ pc^{-1}. If we only consider thermal support against gravity, such a high line-mass structure cannot keep 0.1 pc. Many authors studied the effects of turbulence and/or magnetic field and have shown that sub-critical and mildly super-critical filaments have a width of 0.1 pc (Fischera and Martin 2012; Auddy et al. 2016; Priestley and Whitworth 2022; Federrath 2016; Ntormousi and Hennebelle 2019), but still, the constant width of filaments particularly for massive filaments remains a mystery. Still, we need to understand the origin of universal width particularly for massive filaments by examining the detailed process of gas accretion flows onto the filament. We expect that the SSI can deposit additional energy to the filament, because the corrugation of the shock front generally leads to turbulent flows behind the shock (e.g., Inoue et al. 2012; Inoue and Inutsuka 2012b). Equations (1.22) and (1.23) show that the most unstable scale is infinitesimally small. This unphysical feature is caused by ideal approximation and results in the discontinuous treatment of the shock. To know the physical scale length of the SSI, we take ambipolar diffusion into account. As the first step to understanding the effect of SSI on filament width, we study the effect of ambipolar diffusion on the SSI and derive the most unstable scale.

References

Abe D, Inoue T, Inutsuka S-I (2024) ApJ 961:100. https://doi.org/10.3847/1538-4357/ad072a, https://ui.adsabs.harvard.edu/abs/2024ApJ...961..100A

Abe D, Inoue T, Inutsuka S-I, Matsumoto T (2021) ApJ 916:83. https://doi.org/10.3847/1538-4357/ac07a1, https://ui.adsabs.harvard.edu/abs/2021ApJ...916...83A

André P et al (2010) A&A 518:L102. https://doi.org/10.1051/0004-6361/201014666, https://ui.adsabs.harvard.edu/abs/2010A&A...518L.102A
André P, Arzoumanian D, Könyves V, Shimajiri Y, Palmeirim P (2019) Astron Astrophys 629:1. https://doi.org/10.1051/0004-6361/201935915
André P, Di Francesco J, Ward-Thompson D, Inutsuka SI, Pudritz RE, Pineda JE (2014) In: Beuther H, Klessen RS, Dullemond CP, Henning T (eds) Protostars and Planets VI, p 27
Arzoumanian D et al (2011) A&A 529:L6. https://doi.org/10.1051/0004-6361/201116596, https://ui.adsabs.harvard.edu/abs/2011A&A...529L...6A
Arzoumanian D, André P, Peretto N, Könyves V (2013) A&A 553:A119. https://doi.org/10.1051/0004-6361/201220822, http://adsabs.harvard.edu/abs/2013A%26A...553A.119A
Arzoumanian D, Shimajiri Y, Inutsuka S, Inoue T, Tachihara K (2018) PASJ 70:96. https://doi.org/10.1093/pasj/psy095, https://ui.adsabs.harvard.edu/abs/2018PASJ...70...96A
Auddy S, Basu S, Kudoh T (2016) ApJ 831:46. https://doi.org/10.3847/0004-637X/831/1/46, https://ui.adsabs.harvard.edu/abs/2016ApJ...831...46A
Balfour SK, Whitworth AP, Hubber DA (2017) MNRAS 465:3483. https://doi.org/10.1093/mnras/stw2956, https://ui.adsabs.harvard.edu/abs/2017MNRAS.465.3483B
Balfour S. K., Whitworth AP, Hubber DA, Jaffa SE (2015) MNRAS 453:2471. https://doi.org/10.1093/mnras/stv1772, https://ui.adsabs.harvard.edu/abs/2015MNRAS.453.2471B
Beresnyak A, Lazarian A, Cho J (2005) ApJ 624:L93. https://doi.org/10.1086/430702, https://ui.adsabs.harvard.edu/abs/2005ApJ...624L..93B
Bonne L et al (2020) A&A 644:A27. https://doi.org/10.1051/0004-6361/202038281, https://ui.adsabs.harvard.edu/abs/2020A&A...644A..27B
Chandrasekhar S, Fermi E (1953) ApJ 118:113. https://doi.org/10.1086/145731, https://ui.adsabs.harvard.edu/abs/1953ApJ...118..113C
Chen C-Y, Mundy LG, Ostriker EC, Storm S, Dhabal A (2020) MNRAS 494:3675. https://doi.org/10.1093/mnras/staa960, https://ui.adsabs.harvard.edu/abs/2020MNRAS.494.3675C
Chen CY, Ostriker EC (2014) ApJ 785. https://doi.org/10.1088/0004-637X/785/1/69
Chen C-Y (2015), PhD thesis, pp 1–212
Chira RA, Kainulainen J, Ibáñez-Mejía JC, Henning T, Mac Low MM (2018) A&A 610:A62. https://doi.org/10.1051/0004-6361/201731836, https://ui.adsabs.harvard.edu/abs/2018A&A...610A..62C
Clarke SD, Whitworth AP, Duarte-Cabral A, Hubber DA (2017) MNRAS 468:2489. https://doi.org/10.1093/mnras/stx637, https://ui.adsabs.harvard.edu/abs/2017MNRAS.468.2489C
Clarke SD, Whitworth AP, Hubber DA (2016) MNRAS 458:319. https://doi.org/10.1093/mnras/stw407, https://ui.adsabs.harvard.edu/abs/2016MNRAS.458..319C
Crutcher RM (2012) ARA&A 50:29. https://doi.org/10.1146/annurev-astro-081811-125514, https://ui.adsabs.harvard.edu/abs/2012ARA&A..50...29C
Davis L (1951) Phys Rev 81:890. https://doi.org/10.1103/PhysRev.81.890.2, https://ui.adsabs.harvard.edu/abs/1951PhRv...81..890D
Draine BT, Roberge WG, Dalgarno A (1983) ApJ 264:485. https://doi.org/10.1086/160617, https://ui.adsabs.harvard.edu/abs/1983ApJ...264..485D
Édel'Man MA (1989) Astrophysics 31:758. https://doi.org/10.1007/BF01012737, https://ui.adsabs.harvard.edu/abs/1989Ap.....31..758E
Elmegreen BG, Scalo J (2004) ARA&A 42:211. https://doi.org/10.1146/annurev.astro.41.011802.094859, https://ui.adsabs.harvard.edu/abs/2004ARA&A..42..211E
Enoch ML, Glenn J, Evans Neal JI, Sargent AI, Young KE, Huard TL (2007) ApJ 666:982. https://doi.org/10.1086/520321, https://ui.adsabs.harvard.edu/abs/2007ApJ...666..982E
Enokiya R, Torii K, Fukui Y (2021) PASJ 73:S75. https://doi.org/10.1093/pasj/psz119, https://ui.adsabs.harvard.edu/abs/2021PASJ...73S..75E
Federrath C (2016) MNRAS 457:375. https://doi.org/10.1093/mnras/stv2880, https://ui.adsabs.harvard.edu/abs/2016MNRAS.457..375F
Field GB (1965) ApJ 142:531. https://doi.org/10.1086/148317, https://ui.adsabs.harvard.edu/abs/1965ApJ...142..531F

Fischera J, Martin PG (2012) A&A 542:A77. https://doi.org/10.1051/0004-6361/201218961, https://ui.adsabs.harvard.edu/abs/2012A&A...542A..77F

Fukui Y et al (2019) ApJ 886:14. https://doi.org/10.3847/1538-4357/ab4900, https://ui.adsabs.harvard.edu/abs/2019ApJ...886...14F

Heigl S, Burkert A, Gritschneder M (2018) MNRAS 474:4881. https://doi.org/10.1093/mnras/stx3145, https://ui.adsabs.harvard.edu/abs/2018MNRAS.474.4881H

Heigl S, Gritschneder M, Burkert A (2020) MNRAS 495:758. https://doi.org/10.1093/mnras/staa1202, https://ui.adsabs.harvard.edu/abs/2020MNRAS.495..758H

Hennebelle P (2013) A&A 556:A153. https://doi.org/10.1051/0004-6361/201321292, https://ui.adsabs.harvard.edu/abs/2013A&A...556A.153H

Hennebelle P, Inutsuka S (2019) Front Astron Space Sci 6:5.https://doi.org/10.3389/fspas.2019.00005, https://ui.adsabs.harvard.edu/abs/2019FrASS...6....5H

Henshaw JD, Caselli P, Fontani F, Jiménez-Serra I, Tan JC (2014) MNRAS 440:2860. https://doi.org/10.1093/mnras/stu446, https://ui.adsabs.harvard.edu/abs/2014MNRAS.440.2860H

Inoue T, Fukui Y (2013) ApJ 774:L31. https://doi.org/10.1088/2041-8205/774/2/L31, https://ui.adsabs.harvard.edu/abs/2013ApJ...774L..31I

Inoue T, Hennebelle P, Fukui Y, Matsumoto T, Iwasaki K, Inutsuka S-I (2018) PASJ 70:S53. https://doi.org/10.1093/pasj/psx089, https://ui.adsabs.harvard.edu/abs/2018PASJ...70S..53I

Inoue T, Inutsuka S (2012a) ApJ 759:35. https://doi.org/10.1088/0004-637X/759/1/35, https://ui.adsabs.harvard.edu/abs/2012ApJ...759...35I

Inoue T, Inutsuka S (2016) ApJ 833:10. https://doi.org/10.3847/0004-637X/833/1/10, https://ui.adsabs.harvard.edu/abs/2016ApJ...833...10I

Inoue T, Inutsuka S-I (2009) ApJ 704:161. https://doi.org/10.1088/0004-637X/704/1/161, https://ui.adsabs.harvard.edu/abs/2009ApJ...704..161I

Inoue T, Inutsuka S-I (2012b) ApJ 759:35. https://doi.org/10.1088/0004-637X/759/1/35, https://ui.adsabs.harvard.edu/abs/2012ApJ...759...35I

Inoue T, Yamazaki R, Inutsuka S-I, Fukui Y (2012) ApJ 744:71. https://doi.org/10.1088/0004-637X/744/1/71, https://ui.adsabs.harvard.edu/abs/2012ApJ...744...71I

Inutsuka S-I, Inoue T, Iwasaki K, Hosokawa T (2015) A&A 580:A49. https://doi.org/10.1051/0004-6361/201425584, https://ui.adsabs.harvard.edu/abs/2015A&A...580A..49I

Inutsuka S-I, Miyama SM (1992b) ApJ 388:392. https://doi.org/10.1086/171162, http://adsabs.harvard.edu/abs/1992ApJ...388..392I

Inutsuka S, Miyama SM (1992a) ApJ 388:392. https://doi.org/10.1086/171162, https://ui.adsabs.harvard.edu/abs/1992ApJ...388..392I

Inutsuka S, Miyama SM (1997) ApJ 480:681. https://doi.org/10.1086/303982

Kashiwagi R, Tomisaka K (2021) ApJ 911:106. https://doi.org/10.3847/1538-4357/abea7a, https://ui.adsabs.harvard.edu/abs/2021ApJ...911..106K

Kitsionas S, Whitworth AP (2007) MNRAS 378:507. https://doi.org/10.1111/j.1365-2966.2007.11707.x, https://ui.adsabs.harvard.edu/abs/2007MNRAS.378..507K

Klessen RS, Hennebelle P (2010) A&A 520:A17. https://doi.org/10.1051/0004-6361/200913780, https://ui.adsabs.harvard.edu/abs/2010A&A...520A..17K

Könyves V et al (2015) A&A 584:A91. https://doi.org/10.1051/0004-6361/201525861, https://ui.adsabs.harvard.edu/abs/2015A&A...584A..91K

Koyama H, Inutsuka S (2000) ApJ 538:980. https://doi.org/10.1086/308594

Lada CJ, Lombardi M, Alves JF (2010) ApJ 724:687. https://doi.org/10.1088/0004-637X/724/1/687, https://ui.adsabs.harvard.edu/abs/2010ApJ...724..687L

Larson RB (1981) MNRAS 194:809. https://doi.org/10.1093/mnras/194.4.809, https://ui.adsabs.harvard.edu/abs/1981MNRAS.194..809L

Lessen M, Deshpande NV (1967) J Plasma Phys 1:463. https://doi.org/10.1017/S0022377800003457, https://ui.adsabs.harvard.edu/abs/1967JPlPh...1..463L

Matsumoto T, Dobashi K, Shimoikura T (2015) ApJ 801:77. https://doi.org/10.1088/0004-637X/801/2/77, https://ui.adsabs.harvard.edu/abs/2015ApJ...801...77M

Miyama SM, Narita S, Hayashi C (1987a) PThPh 78:1051. https://doi.org/10.1143/PTP.78.1051, https://ui.adsabs.harvard.edu/abs/1987PThPh..78.1051M
Miyama SM, Narita S, Hayashi C (1987b) PThPh 78:1273. https://doi.org/10.1143/PTP.78.1273, https://ui.adsabs.harvard.edu/abs/1987PThPh..78.1273M
Nagai T, Inutsuka S-I, Miyama SM (1998) ApJ 506:306. https://doi.org/10.1086/306249, https://ui.adsabs.harvard.edu/abs/1998ApJ...506..306N
Ntormousi E, Hennebelle P (2019) A&A 625:A82. https://doi.org/10.1051/0004-6361/201834094, https://ui.adsabs.harvard.edu/abs/2019A&A...625A..82N
Orkisz JH et al (2019) A&A 624:A113. https://doi.org/10.1051/0004-6361/201833410, https://ui.adsabs.harvard.edu/abs/2019A&A...624A.113O
Osterbrock DE (1961) ApJ 134:270. https://doi.org/10.1086/147155, https://ui.adsabs.harvard.edu/abs/1961ApJ...134..270O
Ostriker J (1964) ApJ 140:1056. https://doi.org/10.1086/148005, https://ui.adsabs.harvard.edu/abs/1964ApJ...140.1056O
Padoan P, Nordlund Å (1999) ApJ 526:279. https://doi.org/10.1086/307956, https://ui.adsabs.harvard.edu/abs/1999ApJ...526..279P
Palmeirim P et al (2013) A&A 350:A38. https://doi.org/10.1051/0004-6361/201220500, https://ui.adsabs.harvard.edu/abs/2013A&A...550A..38P
Peretto N et al (2014) A&A 561:A83. https://doi.org/10.1051/0004-6361/201322172, https://ui.adsabs.harvard.edu/abs/2014A&A...561A..83P
Pineda JE et al (2022) https://ui.adsabs.harvard.edu/abs/2022arXiv220503935P, arXiv:2205.03935
Planck Collaboration XXXV (2016) A&A 586:A138. https://doi.org/10.1051/0004-6361/201525896, https://ui.adsabs.harvard.edu/abs/2016A&A...586A.138P
Priestley FD, Whitworth AP (2022) MNRAS 512:1407. https://doi.org/10.1093/mnras/stac627, https://ui.adsabs.harvard.edu/abs/2022MNRAS.512.1407P
Pudritz RE, Kevlahan NKR (2013) Philos Trans R Soc Lond Ser A 371:20120248. https://doi.org/10.1098/rsta.2012.0248, https://ui.adsabs.harvard.edu/abs/2013RSPTA.37120248P
Sakre N, Habe A, Pettitt AR, Okamoto T (2021) PASJ 73:S385. https://doi.org/10.1093/pasj/psaa059, https://ui.adsabs.harvard.edu/abs/2021PASJ...73S.385S
Scalo J, Elmegreen BG (2004) ARA&A 42:275. https://doi.org/10.1146/annurev.astro.42.120403.143327, https://ui.adsabs.harvard.edu/abs/2004ARA&A..42..275S
Seifried D, Walch S (2015) MNRAS 452. https://doi.org/10.1093/mnras/stv1458
Shimajiri Y, André P, Ntormousi E, Men'shchikov A, Arzoumanian D, Palmeirim P (2019) A&A 632:A89. https://doi.org/10.1051/0004-6361/201935689, https://ui.adsabs.harvard.edu/abs/2019A&A...632A..83S
Soler JD, Hennebelle P (2017) A&A 607:A2. https://doi.org/10.1051/0004-6361/201731049, https://ui.adsabs.harvard.edu/abs/2017A&A...607A...2S
Stodólkiewicz JS (1963) Acta Astron 13:30. https://ui.adsabs.harvard.edu/abs/1963AcA....13...30S
Suri S et al (2019) A&A 623:A142. https://doi.org/10.1051/0004-6361/201834049, https://ui.adsabs.harvard.edu/abs/2019A&A...623A.142S
Takahira K, Tasker EJ, Habe A (2014) ApJ 792:63. https://doi.org/10.1088/0004-637X/792/1/63, https://ui.adsabs.harvard.edu/abs/2014ApJ...792...63T
Tokuda K et al (2019) ApJ 886:15. https://doi.org/10.3847/1538-4357/ab48ff, https://ui.adsabs.harvard.edu/abs/2019ApJ...886...15T
Tomisaka K (2014) ApJ 785. https://doi.org/10.1088/0004-637X/785/1/24
Tomisaka K, Ikeuchi S (1983) PASJ 35:187. https://ui.adsabs.harvard.edu/abs/1983PASJ...35..187T
Vaidya B, Hartquist TW, Falle SAEG (2013) MNRAS 433:1258. https://doi.org/10.1093/mnras/stt800, https://ui.adsabs.harvard.edu/abs/2013MNRAS.433.1258V
Wolfire MG, Hollenbach D, McKee CF, Tielens AGGM, Bakes ELO (1995) ApJ 443:152. https://doi.org/10.1086/175510, https://ui.adsabs.harvard.edu/abs/1995ApJ...443..152W
Xu S, Ji S, Lazarian A (2019) ApJ 878:157. https://doi.org/10.3847/1538-4357/ab21be, https://ui.adsabs.harvard.edu/abs/2019ApJ...878..157X

Zucker C et al (2022) Nature 601:334. https://doi.org/10.1038/s41586-021-04286-5, https://ui.adsabs.harvard.edu/abs/2022Natur.601..334Z

Zuckerman B, Evans NJI (1974) ApJ 192:L149. https://doi.org/10.1086/181613, https://ui.adsabs.harvard.edu/abs/1974ApJ...192L.149Z

Chapter 2
Classification of Filament Formation Mechanisms

2.1 Short Introduction

Previous theoretical research exploring filament formation has introduced several mechanisms, as outlined in Sect. 1.2.2. Despite this breadth of proposed mechanisms, the specific type accountable for generating star-forming filaments remains ambiguous and unresolved. Shock compression triggers most proposed mechanisms including types I, O, C, and possibly type G. This study conducts a comprehensive exploration through a series of isothermal MHD simulations. Our primary aim is to investigate filament formation, specifically focusing on the influences of shock strength, turbulence, and self-gravity on the mechanisms governing filament formation. Note that, since type I is effective in highly super-Alfvénic turbulence, we hardly observe it in our simulation results performed under a realistic 10 μG initial magnetic field. Furthermore, we analyze the line-mass distribution of the simulated filaments. The paper is organized as follows: Sect. 2.2 outlines the setup of our simulations, while Sect. 2.3 presents and interprets the results, specifically addressing the filament line-mass distribution. Section 2.4 introduces simple models elucidating the dominant timescale governing filament formation. Finally, Sect. 2.5 provides a comprehensive summary of our findings.

2.2 Setup of Simulations

We use the SFUMATO code (Matsumoto 2007) to solve the isothermal MHD equations, incorporating self-gravity. This code integrates the MHD equations using a Godunov-type scheme and an HLLD Riemann solver (Miyoshi and Kusano 2005), providing third- and second-order accuracy in space and time, respectively. Maintaining the divergence-free state of the magnetic field ($\nabla \cdot \mathbf{B} = 0$) is achieved through

D. Abe, *Formation and Evolution of Star-Forming Filaments in Molecular Clouds*, Springer Theses, https://doi.org/10.1007/978-981-97-6414-3_2

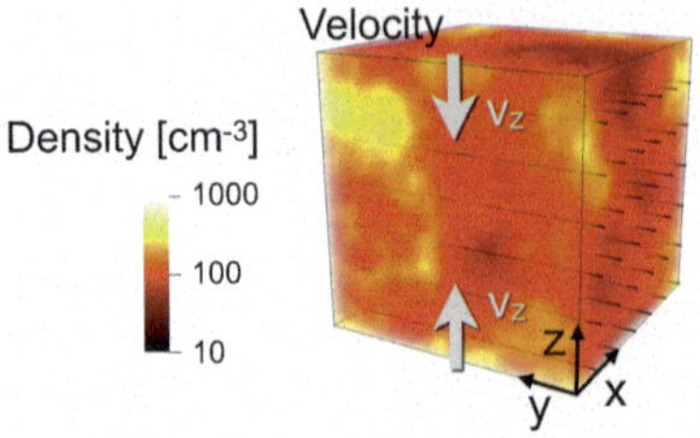

Fig. 2.1 Schematic diagram of initial condition. The color bar corresponds to the magnitude of density, while the black lines denote the initial magnetic field configuration and the grey arrows indicate the orientations of the converging flows. Reproduced from Abe et al. (2021). © 2021. The Author(s). Published by the American Astronomical Society. CC BY 4.0

the divergence cleaning method (Dedner et al. 2002). We use a multi-grid method to solve Poisson's equation.

We investigate the mechanism of filament formation induced by shock waves by performing converging flow simulations. Figure 2.1 presents a schematic illustration showing the initial condition for this investigation. We construct a cubic numerical domain with side lengths $L_{\rm box} = 6.0$ pc, comprising 512^3 uniformly distributed numerical cells. This configuration implies a physical resolution of approximately $\Delta x \approx 0.012$ pc. Initially, we set the isothermal gas characterized by the isothermal sound speed at $c_s = 0.2$ km s^{-1}. Given the inherent highly inhomogeneous nature of molecular clouds, we introduce isotropic density fluctuations. These fluctuations are created through a superposition of sinusoidal functions encompassing various wavenumbers ranging from $2\pi/L_{\rm box}$ to $32\pi/L_{\rm box}$, along with random phases. The density fluctuations follow a power spectrum described by $(\log\rho)_k^2 \propto k^{-4}$, a characteristic anticipated in supersonic turbulence scenarios (Beresnyak et al. 2005; Elmegreen and Scalo 2004; Scalo and Elmegreen 2004; Larson 1981; Heyer and Brunt 2004). Our simulations' initial density structure is parameterized by a mean density $\bar{n}_0 = \bar{\rho}_0/m = 100$ cm^{-3} and a dispersion $\Delta n/\bar{n}_0 = 0.5$, where $\bar{\rho}0$ signifies the mean mass density, and $m = 2.4\,m$proton represents the mean mass of the molecular gas particles. Furthermore, we also set the initial turbulent velocity field based on the model outlined in Table 3.1. The turbulent velocity fluctuation has a dispersion of 1.0 km s^{-1} with a power spectrum of $v_k^2 \propto k^{-4}$, following Larson's law (Larson 1981).

Additionally, we introduce an initial coherent velocity component characterized by $v_z(z) = -(v_{\rm coll}/2)\tanh[z-3]$. This setting represents two flows colliding head-on within the x-y plane at $z = 3$ pc. Previous studies highlighting type C filament formation observed filaments emerging within the shock-compressed layer, characterized by shock velocities of a few km s^{-1}. Conversely, investigations into the type O mechanism focused on cases with high shock velocities, approximately ~ 10 km s^{-1}. To comprehensively investigate the filament formation mechanism, we perform simulations across a range of shock velocities. Specifically, we explore cases with $v_{\rm coll} = 3$, $v_{\rm coll} = 6$, $v_{\rm coll} = 8$, $v_{\rm coll} = 10$, and 12 km s^{-1}. A summary of

Table 2.1 Model parameters

Model name	Collision velocity $v_{\rm coll}$ [km s^{-1}]	Shock velocity $v_{\rm sh}$[km s^{-1}]	Self-gravity	Turbulence
v12GyTn	12	7.0	Yes	No
v12GnTn	12	7.0	No	No
v12GyTy	12	7.0	Yes	Yes
v12GnTy	12	7.0	No	Yes
v10GyTn	10	6.0	Yes	No
v8GyTn	8.0	5.0	Yes	No
v6GyTn	6.0	4.0	Yes	No
v3GyTn	3.0	2.5	Yes	No
v3GnTn	3.0	2.5	No	No
v3GyTy	3.0	2.5	Yes	Yes
v3GnTy	3.0	2.5	No	Yes

the model parameters is provided in Table 2.1. Within the model names, the numeric values following "v" denote the respective $v_{\rm coll}$ values in units of km s^{-1}. The characters "y" and "n" following "G" and "T" represent simulation with and without self-gravity and initial turbulence, respectively.

At the x-y boundary planes, the velocity is held constant at $v_{\rm coll}/2$, while the density is specified as $n_0(x, y, z = v_{\rm coll}t/2)$ for the $z = 6$ pc plane and $n_0(x, y, z = L_{\rm box} - v_{\rm coll}t/2)$ for the $z = 0$ pc plane, where $n_0(x, y, z)$ represents the initial density field incorporating fluctuations. Regarding the z-x and y-z boundary planes, we use periodic boundary conditions.

Initially, we set a uniform magnetic field of $\mathbf{B}_0 = (0, 10\,\mu\mathrm{G}, 0)$, oriented perpendicular to the direction of the shock propagation. This magnetic field strength aligns with observed magnitudes found in molecular clouds (e.g., Crutcher 2012; Heiles and Crutcher 2005). The magnetic field component perpendicular to the direction of the shock propagation is anticipated to amplify due to shock compression, while the parallel component is not expected to experience such amplification. Consequently, even if an initial z-component of the magnetic field is provided, its influence is projected to be minor in the dynamics.

We perform simulations both with and without considering self-gravity to investigate its impact on filament formation. When self-gravity is incorporated, we adopt the sink particle technique within regions where gravitational collapse is anticipated to occur. The criteria for generating sink particles is the same as previously established conditions, as in prior research (Inoue et al. 2018; Matsumoto et al. 2015). It is noted that the use of sink particles is primarily to enable the continuation of simulations even after the onset of local gravitational collapse[1]. In this paper, we

[1] The threshold density for sink formation is 5.6×10^4 cm^{-3}. This value is notably lower than thresholds employed in prior studies due to the absence of adaptive mesh refinement in this study.

will not focus extensively on sink particle information. This is because the cumulative mass attributed to sink particles is considerably smaller than that encapsulated within dense filaments. For instance, in model v12GyTy, the final total mass of sink particles accounts for only 3% compared to that of dense filaments, while for model v3GyTy, it constitutes 6% .

2.3 Results

2.3.1 *High Shock Velocity Case*

For $v_{\rm coll} = 12$ km s^{-1}, the collision of flows generates fast isothermal MHD shock waves. According to the shock jump condition (e.g., Sect. 4.1 of Fukui et al. 2020), the compression ratio of the isothermal fast shock is given by

$$\rho_1/\rho_0 = v_0/v_1 \simeq \sqrt{2}\mathcal{M}_{\rm A}, \tag{2.1}$$

where $\mathcal{M}_A \equiv v_0/v_{\rm Alf,0}$ is the Alfvénic Mach number, the Alfvén velocity is given by $v_{\rm Alf,0} = B_0/\sqrt{4\pi\rho_0}$ and subscripts 0 and 1 indicate preshock and postshock values respectively. The velocity v_0, observed in the shock rest-frame, is equivalent to the shock velocity in the upstream rest frame, denoted as $v_{\rm sh}$. As the numerical domain is within the postshock rest frame, the shock wave propagates with a velocity approximately defined as $v_1 \simeq v_0/(\sqrt{2}\mathcal{M}_{\rm A})$. The relation between the converging flow velocity $v_{\rm coll}$ and the shock velocity $v_{\rm sh}$ is given by

$$\begin{aligned} v_{\rm sh} &= v_{\rm coll}/2 + v_1 \\ &\simeq v_{\rm coll}/2 + v_{\rm Alf,0}/\sqrt{2} \\ &\simeq 6\left(\frac{v_{\rm coll}}{12\ {\rm km\ s^{-1}}}\right)\ {\rm km\ s^{-1}} \\ &\quad + 1\left(\frac{B_0}{10\ \mu{\rm G}}\right)\left(\frac{n_0}{100\ {\rm cm^{-3}}}\right)^{-1/2}\ {\rm km\ s^{-1}}. \end{aligned} \tag{2.2}$$

In Fig. 2.2, we present snapshots illustrating the column density maps derived from the results obtained in models v12GyTn and v12GnTn at $t = 0.2$ (top), 0.40 (middle), and 0.60 (bottom) Myr. When constructing the column density map, we specifically identify the shock fronts and compute the density within the shock-compressed region. To identify the shock fronts, a procedure is adopted where the total pressure, $\rho(x, y, z)c_s^2 + B^2(x, y, z)/8\pi$, is scanned along the z-axis from upstream ($z = 0$ and 6) to downstream ($z = 3$). The two shock fronts are defined based on the largest and smallest positions of $z(x, y)$ satisfying the following condition:

$$\rho(x, y, z)c_s^2 + B^2(x, y, z)/8\pi \geq f_{\rm jump} \times \bar{\rho}_0 v_{\rm coll}^2. \tag{2.3}$$

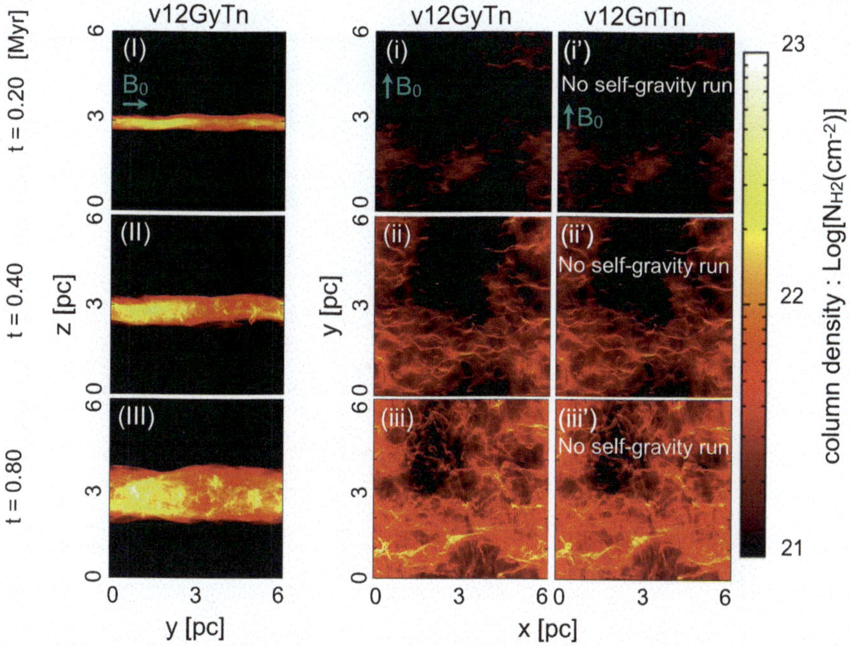

Fig. 2.2 Column density maps at time $t = 0.2$ (*top*), 0.4 (*middle*), and 0.8 (*bottom*) Myr. *Left row* (panels I, II, and III): Column density in the *y*-*z* plane of model v12GyTn. *Middle row* (panels i, ii, and iii): Same as panels (I)–(III) but for the *x*-*y* plane. *Right row* (panels i', ii', and iii'): Same as panels (i)–(iii) but for model v12GnTn. Reproduced from Abe et al. (2021).

We selected $f_{\rm jump} = 0.2$ for convenience. However, it is noted that we verified the similarity of results obtained with $f_{\rm jump} = 0.4$, confirming their indistinguishable nature compared to the $f_{\rm jump} = 0.2$ case.

In Fig. 2.2, panels (I)–(III) and (i)–(iii) represent column density snapshots of model v12GyTn, showcasing the *y*-*z* and *x*-*y* planes, respectively. Panels (I), (II), and (III) show two shock waves initiated by converging flows, propagating towards the positive and negative *z*-directions. Panels (i)–(iii) and (i')–(iii') show the emergence of filaments, irrespective of the influence of self-gravity. This suggests that filament formation in this high shock velocity case is not primarily driven by self-gravity. In Sect. 2.3.4, we present that the formation of exceptionally massive filaments reaching 100 $M_\odot$ pc^{-1} at $t = 1$ Myr in this high shock velocity result. Given that the present models (v12GyTn and v12GnTn) do not incorporate initial turbulence, the evidence strongly supports the notion that the filament formation is conclusively attributed to the type O mechanism.

In Fig. 2.3, snapshots at $t = 0.20$ (top), 0.40 (middle), and 0.80 (bottom) Myr from models v12GyTy (panels I–III and i–iii) and v12GnTy (panels i'–iii') are presented. In panels (i)–(iii) and (i')–(iii'), an increased presence of more predominantly faint

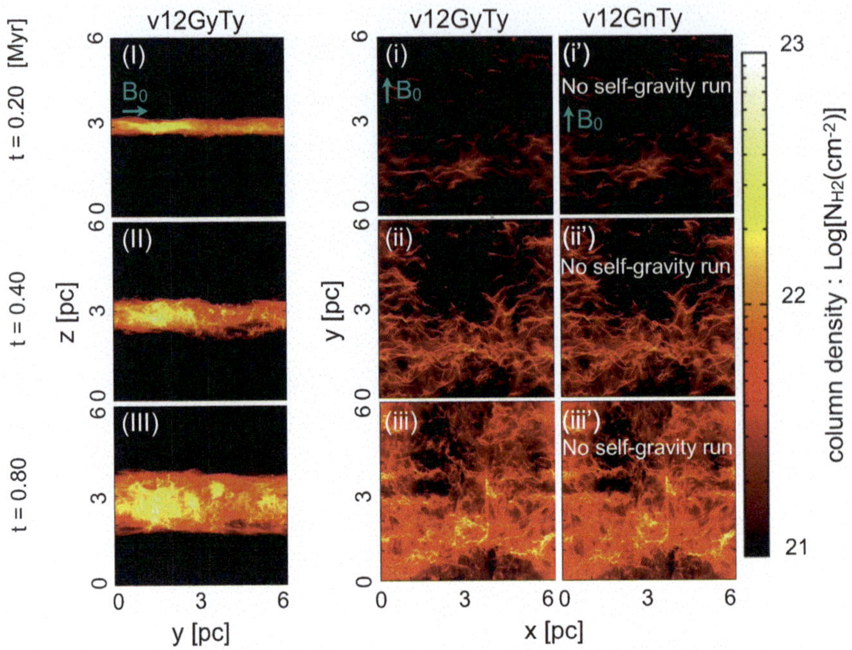

Fig. 2.3 Column density maps at time $t = 0.2$ (*top*), 0.4 (*middle*), and 0.8 (*bottom*) Myr. *Left row* (panels I, II, and III): Column density in the y-z plane of model v12GyTy. *Middle row* (panels i, ii, and iii): Same as panels (I)–(III) but for the x-y plane. *Right row* (panels i', ii', and iii'): Same as panels (i)–(iii) but for model v12GnTy. Reproduced from Abe et al. (2021).

filaments oriented parallel to the mean magnetic field lines is noticeable, compared to the scenario depicted in Fig. 2.2. This suggests the involvement of the type S mechanism in generating these faint filaments. Upon comparison between these models, it is reaffirmed that type G does not contribute significantly to filament formation. However, as shown in Fig. 2.8 (to be discussed below), the resulting filaments tend to disperse in the absence of self-gravity.

In Figs. 2.4 and 2.5, we present local density cross-sections encompassing the five principal filaments, derived from the results of models v12GyTn and v12GyTy, respectively. The highlighted high-density regions (depicted in yellow) within these local cross-section maps correspond to the major filament structures. In the context of the type O mechanism, the post-shock gas moves towards an area situated behind the concave shock surface, facilitating the formation of filaments within this region. The distinct flow pattern emerges as a consequence of the effect of the oblique MHD shock wave. Details into this flow structure have been reported in studies by Inoue and Fukui (2013) and Inoue et al. (2018). The curved morphology of the shock and the velocity vectors (depicted as black arrows) are evident in the cross-section panels

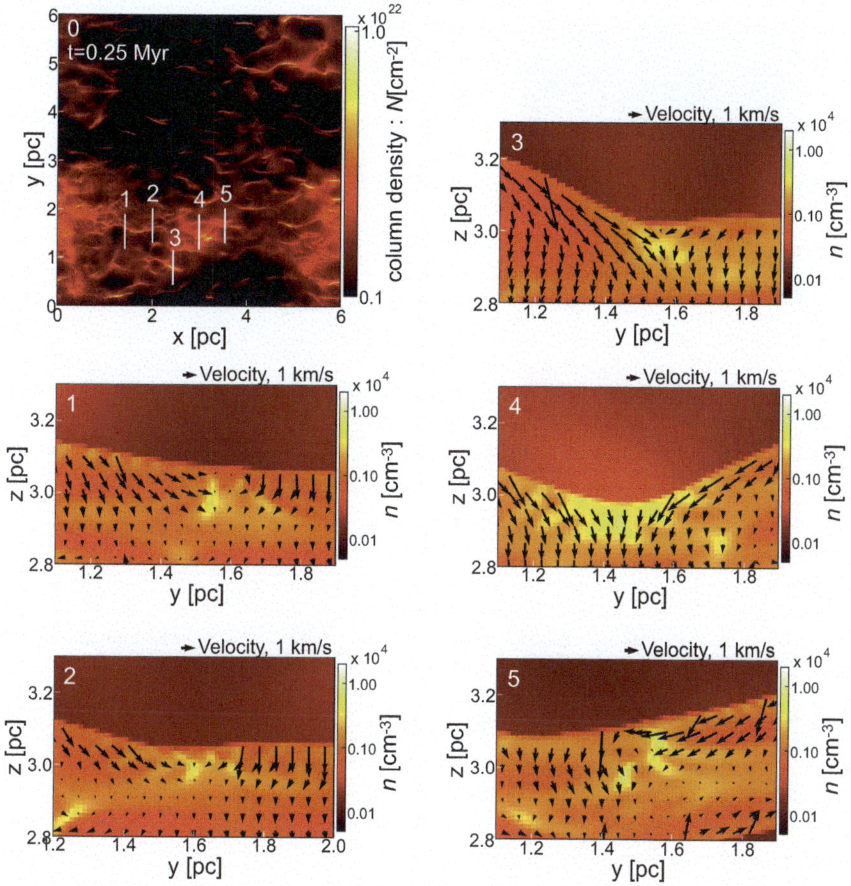

Fig. 2.4 Panel (0): Early stage ($t = 0.25$ Myr) column density map in the x-y plane of result of model v12GyTn. The five white lines mark the planes in which the cross-section maps in panels (1)–(5) are drawn. Panels (1)–(5): Cross-section maps of the number density in the y-z plane. The yellow blobs located roughly at the center of each panel correspond to cross-sections of the filaments formed by the type O mechanism. Reproduced from Abe et al. (2021). © 2021. The Author(s). Published by the American Astronomical Society. CC BY 4.0

within both the v12GyTn and v12GyTy models provide substantial support for the type O origin of the (major) filaments.

The flows generated within the post-shock region resulting from the oblique MHD shock have the potential to induce turbulence in the long-term evolution of the system. However, particularly during the filament formation stage, the converging flows aligned with the bending magnetic field exhibit characteristics of laminar flow. Hence, we highlight that attributing type O filament formation to general supersonic turbulence simulations might not be appropriate, considering the specific stage and dynamics involved in the formation process of these filaments. Indeed, the type O

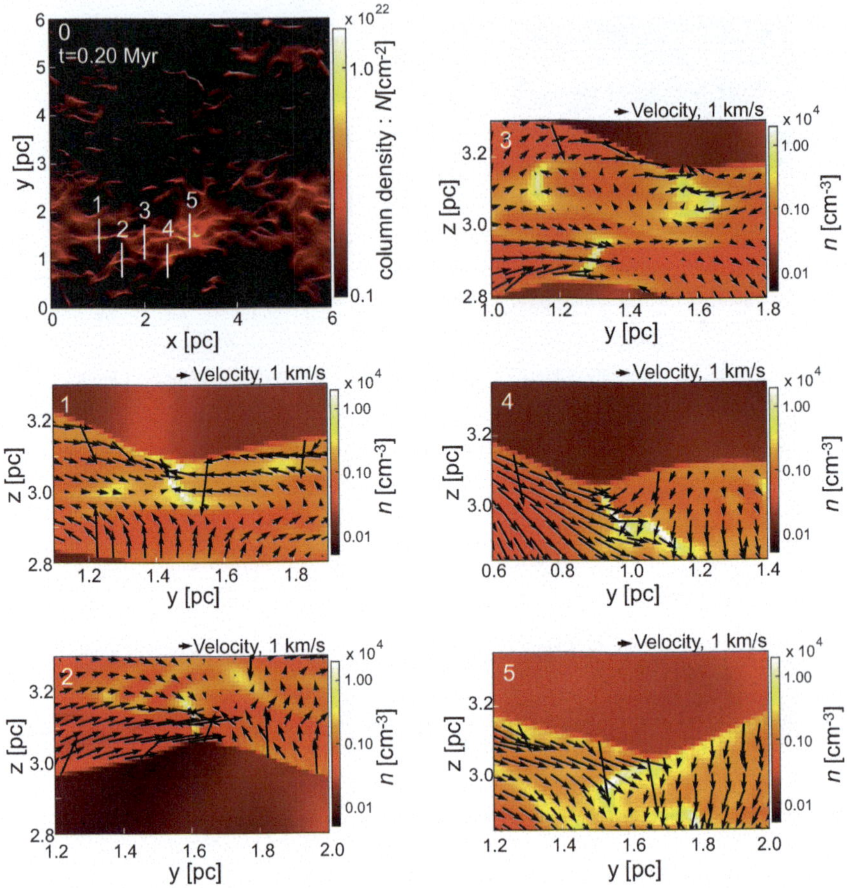

Fig. 2.5 Same as Fig. 2.4 but for model v12GyTy. We can confirm that the type O mechanism takes a major role in filament formation even in the case of initial turbulence. Reproduced from Abe et al. (2021).

mechanism tends to manifest selectively in compressions induced by shock waves characterized by relatively smooth surfaces. Instances of this could arise from the expansion of a super-shell or HII region. It is essential to note that observationally discerning the type O mechanism from other mechanisms can be achieved by scrutinizing the structural attributes of magnetic and velocity fields. This distinction has been explored in observational studies by researchers such as Arzoumanian et al. (2018), Tahani et al. (2018, 2019), Bonne et al. (2020), Chen et al. (2020), Kandori et al. (2020a, b).

2.3.2 Low Shock Velocity Case

In Fig. 2.6, snapshots at $t = 0.6$ (top), 1.4 (middle), and 1.8 (bottom) Myr are presented for models v3GyTn (panels I–III and i–iii) and v3GnTn (panels i'–iii'). Based on Eq. (2.2), the average shock velocity for the $v_{\rm coll} = 3$ km s^{-1} simulations in this section is calculated to be 2.5 km s^{-1}. While filamentary structures are discernible in panel (iii), no distinct dense filaments are evident in panel (iii'). This suggests that the type O mechanism might not be operational in this scenario of low shock velocity. Instead, the filament formation in model v3GyTn is likely attributable to self-gravity (type G).

The snapshots in Fig. 2.7 at $t = 0.4$ (top), 1.20 (middle), and 2.00 (bottom) Myr show the results of model v3GyTy (panels I–III and i–iii) and model v3GnTy (panels i'–iii'). Observing similar patterns of filamentary structure formation at both $t = 0.4$ and 1.2 Myr across both models strongly suggests the significance of type C filament formation specifically in the low shock velocity models.[2] The results from model v3GyTy at $t = 2.0$ Myr indicate a trend wherein filaments are attracted to each other

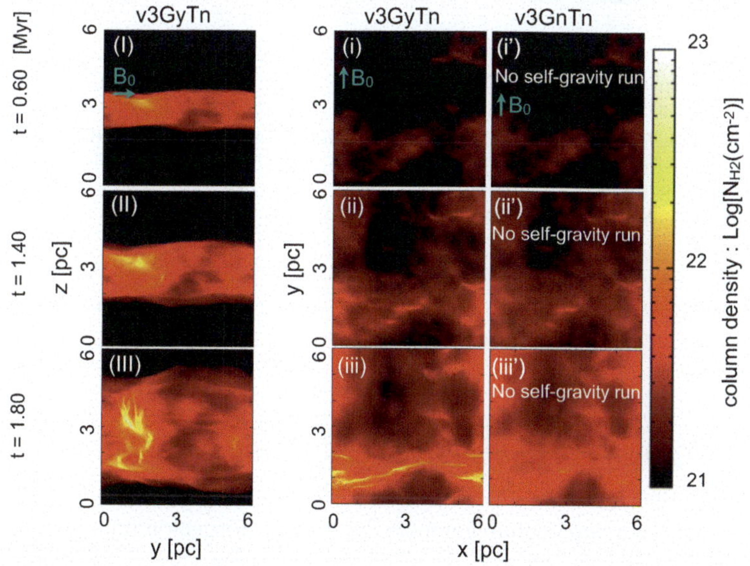

Fig. 2.6 Column density maps at time $t = 0.6$ (*top*), 1.4 (*middle*), and 1.8 (*bottom*) Myr. *Left row* (panels I, II, and III): Column density in the y-z plane of model v3GyTn. *Middle row* (panels i, ii, and iii): Same as the panels (I)–(III) but for the x-y plane. *Right row* (panels i', ii', and iii'): Same as the panels (i)–(iii) but for model v3GnTn. Reproduced from Abe et al. (2021).

[2] Even in the low shock velocity models, some major filaments appear to be formed by type O mechanism.

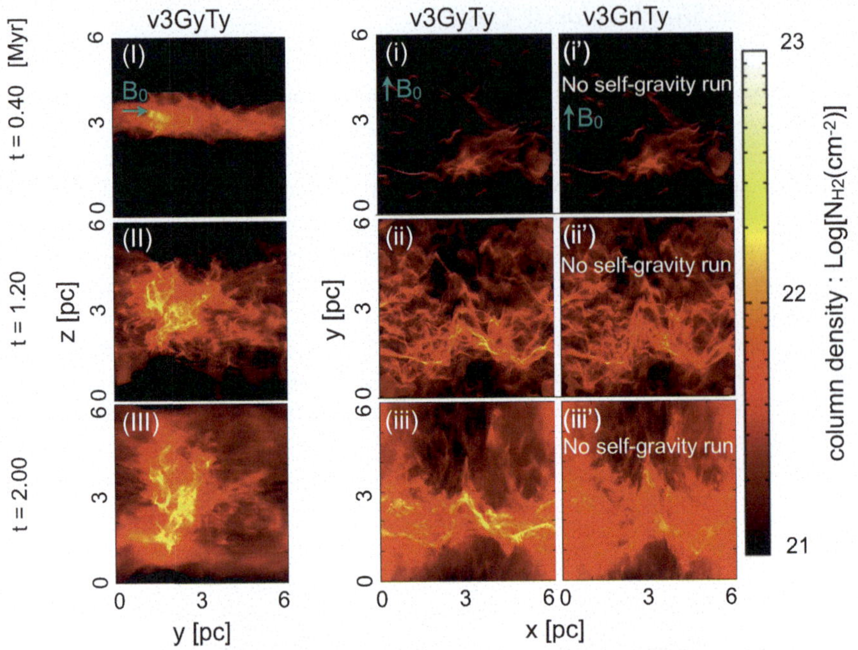

Fig. 2.7 Column density maps at time $t = 0.4$ (*top*), 1.2 (*middle*), and 2.0 (*bottom*) Myr. *Left row* (panels I, II, and III): Column density in the *y*-*z* plane of model v3GyTy. *Middle row* (panels i, ii, and iii): Same as panels (I)–(III) but for the *x*-*y* plane. *Right row* (panels i', ii', and iii'): Same as panels (i)–(iii) but for model v3GnTy. Reproduced from Abe et al. (2021). © 2021. The Author(s). Published by the American Astronomical Society. CC BY 4.0

due to self-gravity, resulting in filament collisions and enhancement of the filament line mass. Additional details are provided in Sect. 2.3.4.

2.3.3 Filament Formation Timescale Versus Free-Fall Time

In this section, our approach to compute the filament formation time involves several steps. Initially, we employ the FilFinder algorithm (Koch and Rosolowsky 2015) to identify filaments, generating filament skeletons based on a provided two-dimensional image. For the analysis, we focus on major filaments while disregarding those with column densities below $1.5 \times \bar{N}_{\rm sh}$, where $\bar{N}_{\rm sh}$ represents the mean column density within the shocked region. We confirm that altering this factor from 1.5 to 2.0 does not affect the results. It is crucial to note that due to this imposed minimum column density requirement for filament identification, our subsequent analysis consistently excludes faint filaments with column densities lower than $1.5 \times \bar{N}_{\rm sh}$.

Then, we calculate the filament mass fraction $F_{\rm fil}$ defined by

$$F_{\rm fil} \equiv M_{\rm fil,tot}/M_{\rm sh}, \tag{2.4}$$

where $M_{\rm fil,tot}$ and $M_{\rm sh}$ are the total mass of the filaments in the snapshot and the shocked region mass, respectively. The total mass $M_{\rm fil,tot}$ is derived by integrating the gas column density over the region approximately 0.1 pc around the filament skeleton (Arzoumanian et al. 2011; Koch and Rosolowsky 2015). To define the filament formation time $t_{\rm fil}$, we use $F_{\rm fil}$, defining it as the period during which filaments are generated most actively. We identify $t_{\rm fil}$ as the time when the rate of change $\Delta F_{\rm fil}/\Delta t$ reaches its maximum value, where Δt and $\Delta F_{\rm fil}$ represent one-tenth of the free-fall time $t_{\rm ff}$ within the shocked layer and the increment of $F_{\rm fil}$ within the time interval Δt, respectively.

The free-fall time in the post-shock layer, which gives the timescale of the self-gravitating sheet fragmentation (Nagai et al. 1998), is estimated as

$$\begin{aligned} t_{\rm ff} &= \sqrt{\frac{1}{2\pi G\bar{\rho}_1}} \\ &= \sqrt{\frac{\bar{v}_{\rm Alf}}{2\sqrt{2}\pi G\bar{\rho}_0\bar{v}_{\rm sh}}} \\ &= \sqrt{\frac{B_0}{4\sqrt{2}\pi^{3/2}G\bar{\rho}_0^{-3/2}\left(v_{\rm coll}/2 + B_0/\sqrt{8\pi\bar{\rho}_0}\right)}} \\ &\simeq 1.0\ {\rm Myr}\left(\frac{B_0}{10\ \mu{\rm G}}\right)^{1/2}\left(\frac{\bar{n}_0}{100\ {\rm cm^{-3}}}\right)^{-3/4} \\ &\times\left[\left(\frac{v_{\rm coll}}{12\ {\rm km/s}}\right) + 0.17\left(\frac{B_0}{10\ \mu{\rm G}}\right)\left(\frac{\bar{n}_0}{100\ {\rm cm^{-3}}}\right)^{-1/2}\right]^{-1/2}, \end{aligned} \tag{2.5}$$

where $\bar{\rho}_1 \simeq \sqrt{2}\mathcal{M}_{\rm A}\bar{\rho}_0$ and $\bar{v}_{\rm Alf} = B_0/\sqrt{4\pi\bar{\rho}_0}$ are the mean density of the shocked layer and the mean Alfvén velocity, respectively (from Eq. 2.1), and $\bar{v}_{\rm sh} = v_{\rm coll}/2 + v_1 \simeq v_{\rm coll}/2 + B_0/\sqrt{8\pi\bar{\rho}_0}$ represents the mean shock velocity (Eq. 2.2).

Figure 2.8 represents the evolution of the filament mass fraction concerning the high shock velocity models, with time normalized by the free-fall time $t_{\rm ff}$. As highlighted in Sect. 2.3.1, the significant filaments in these models originate from the type O mechanism. This figure shows that the timescale for filament formation through the type O mechanism is significantly faster than the timescale of self-gravity.

Figure 2.9 demonstrates the time evolution of the filament mass fraction in the low shock velocity models. In the results obtained from model v3GyTn (panel c), the filament formation time $t_{\rm fil}$ coincides with the free-fall time $t_{\rm ff} \simeq 1.6$ Myr. This supports that the filaments are primarily formed via the type G mechanism as discussed in Sect. 2.3.2. Conversely, in panels (d) and (d'), the emergence of type C occurs earlier than the free-fall time.

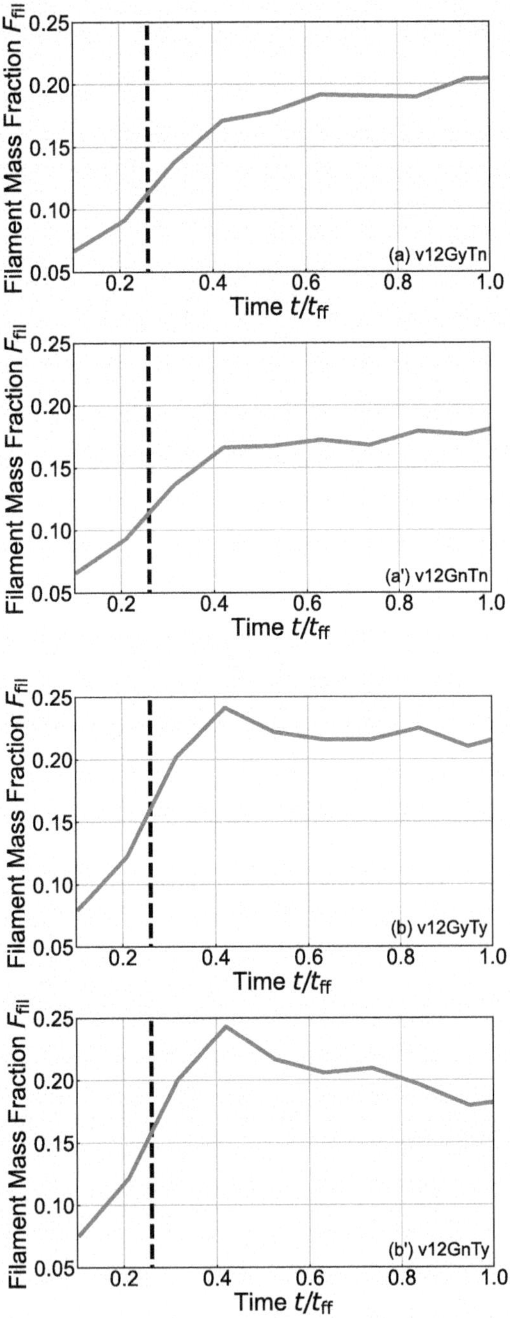

Fig. 2.8 Temporal evolution of the filament mass fraction F_{fil} (solid line) and the filament formation time t_{fil} (dashed line). Panel (a), (a'), (b), and (b') are the results of models v12GyTn, v12GnTn, v12GyTy, and v12GnTy, respectively. Reproduced from Abe et al. (2021).

Fig. 2.9 Temporal evolution of the filament mass fraction F_{fil} (solid line) and the filament formation time t_{fil} (dashed line). Panels (c), (c'), (d), and (d') are the results of models v3GyTn, v3GnTn, v3GyTy, and v3GnTy, respectively. Reproduced from Abe et al. (2021).

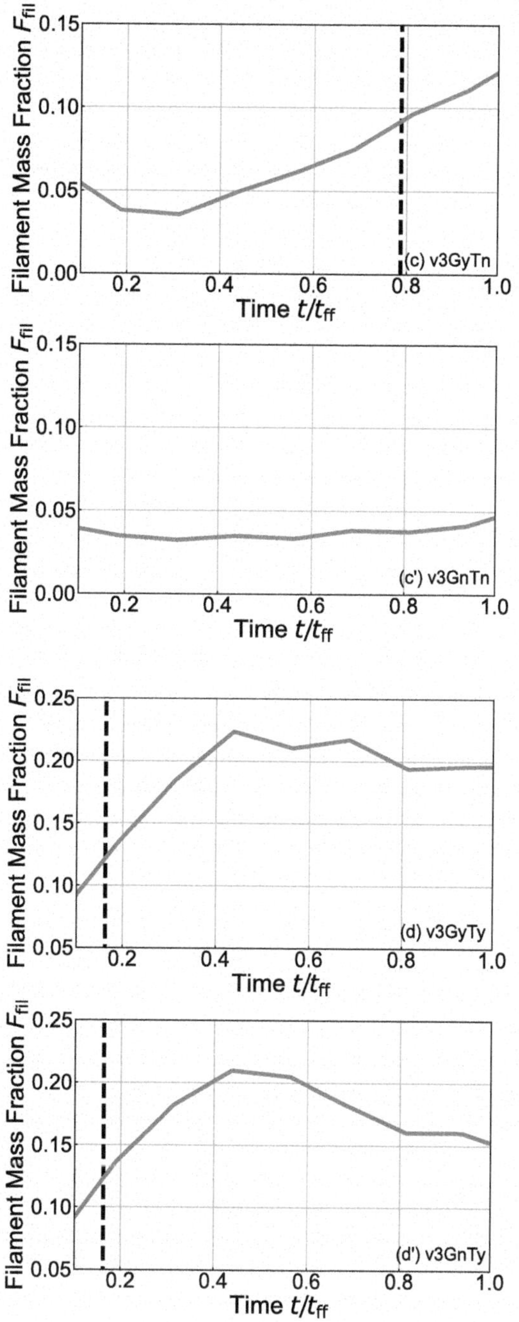

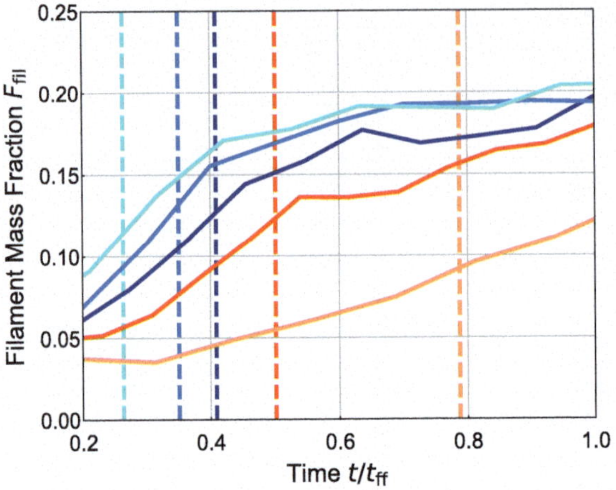

Fig. 2.10 Temporal evolution of the filament mass fraction F_{fil} (solid lines) and the filament formation time t_{fil} (dashed lines) in models with self-gravity and various shock velocities but without turbulent velocity fluctuation. The colors show the results for models v12GyTn (cyan), v10GyTn (blue), v8GyTn (purple), v6GyTn (red), and v3GyTn (orange). Reproduced from Abe et al. (2021).

Figure 2.10 represents the time evolution of the filament mass fraction (solid lines) with the filament formation time (dashed lines) across various shock velocity models that lack turbulent velocity fluctuations, including v12GyTn (cyan), v10GyTn (blue), v8GyTn (purple), v6GyTn (red), and v3GyTn (orange). This shows a clear trend: as shock velocity decreases, the filament formation time proportionally increases. This result indicates a gradual transition in the dominant filament formation mechanism, shifting from the type O mechanism towards the type G mechanism with decreasing shock velocity. Moreover, Fig. 2.10 hints at a threshold collision velocity for the type O mechanism, estimated to be around $v_{coll} \sim 4.5$ km s^{-1}, which corresponds to $v_{sh} \sim 3.3$ km s^{-1}.

2.3.4 *Filament Line Mass Function*

In computing the line-mass histogram, we use the filament skeletons identified in the previous section. Here is how it unfolds: initially, we identify the direction perpendicular to the filament at each grid point on the filament skeleton in the projection plane. Subsequently, the line mass of the filament at each skeleton grid is computed by integrating the gas column density along the normal directions, considering a range of 0.1 pc around the skeleton grid. It is vital to note that the line mass computed in our histogram signifies the local apparent line mass at each skeleton grid

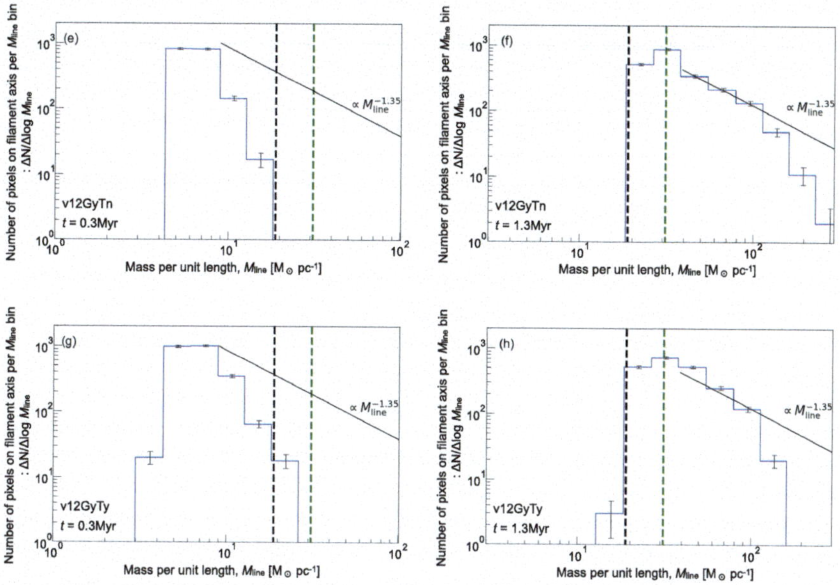

Fig. 2.11 Filament mass functions for model v12GyTn (*top panels* (e) and (f)) and v12GyTy (*bottom panels* (g) and (h)). *Left panels* (e) and (g): Filaments mass functions at time $t = t_{\rm fil}$. *Right panels* (f) and (h): Results at time $t >$ 1 Myr. The black solid line represents the power-law function of the Salpeter initial mass function; the black dashed line shows the critical line-mass (Stodólkiewicz 1963; Ostriker 1964); and the green dashed line is the critical line-mass considering the magnetic field (Tomisaka 2014). Reproduced from Abe et al. (2021). © 2021. The Author(s). Published by the American Astronomical Society. CC BY 4.0

within the 2D projection plane. This measurement is not an average line mass across each filament but rather reflects the localized line mass at specific points along the filament skeleton in the projection plane.

In the top panels of Fig. 2.11, the line-mass functions of model v12GyTn are plotted at $t \simeq t_{\rm fil}$ (top-left) and a later stage of $t = 1.3$ Myr (top-right). The bottom panels are the same but for model v12GyTy. The horizontal axis signifies the local line mass of the filament at a skeleton pixel or a point along the filament axis, denoted as $M_{\rm line}$. The vertical axis represents the count of pixels on the filament axes (skeletons). The pixel size corresponds to 6/512 pc, approximately equaling 0.012 pc. The black line corresponds to the power-law function characterized by the Salpeter index. The black and green dashed lines indicate the critical line mass values, with the former not considering magnetic field support and the latter accounting for magnetic field support, i.e., $M_{\rm line,cr} = 2c_{\rm s}^2/G \sim 17\ {\rm M}_\odot\ {\rm pc}^{-1}$ (Stodólkiewicz 1963; Ostriker 1964) and

$$
\begin{aligned}
M_{\rm line,cr,B} =& 2.24\frac{B_1 w}{G^{1/2}} + 15.4\frac{c_s^2}{G} \\
\simeq & 13.5\left(\frac{w}{0.1\ \rm pc}\right)\left(\frac{\bar{n}_0}{100\ \rm cm^{-3}}\right)^{1/2} \\
&\times\left[\left(\frac{v_{\rm coll}}{12\ \rm km\ s^{-1}}\right)\right. \\
&\left. + 0.17\left(\frac{B_0}{10\ \mu\rm G}\right)\left(\frac{\bar{n}_0}{100\ \rm cm^{-3}}\right)^{-1/2}\right] M_\odot \rm pc^{-1} \\
&+ 15.4\left(\frac{c_s}{0.2\rm km\ s^{-1}}\right)^2 M_\odot \rm pc^{-1},
\end{aligned}
\tag{2.6}
$$

derived by Tomisaka (2014), where $B_1 = \sqrt{2}\mathcal{M}_{\rm A} B_0$ and w are the mean magnetic field in the shocked region and the width of filaments, respectively. Equation (2.6) is derived based on the mean magnetic field strength within the shocked layer, providing an estimation that does not reflect the magnetic field strength specifically threading the filaments. In models v12GyTn and v12GyTy, the results indicate that the majority of line masses at $t \simeq t_{\rm fil}$ (panels [e] and [g] in Fig. 2.11) remain sub-critical. However, these line masses transition into super-critical states within roughly 1 Myr, driven by continuous accretion flows induced by the oblique shock (panels (f) and (h) in Fig. 2.11).

Panels (f) and (h) of Fig. 2.11 represent filament line-mass functions that exhibit a Salpeter-like slope, roughly proportional to $M_{\rm line}^{-1.35}$, particularly at larger line masses. Remarkably, this slope closely resembles the line-mass function discovered in the work of André et al. (2019). This slope is similar to the high-mass part of the core mass function identified by Inutsuka (2001). Understanding the underlying physics contributing to this agreement between specific models and the Salpeter-like slope requires further investigation. This entails conducting additional simulations that involve varying parameters in our models. This will be done in our future work.

Recent observations highlight a significant finding: high line-mass filaments greater than 100 $M_\odot$ pc^{-1} emerge as compelling candidates for potential massive star progenitors (Fukui et al. 2019; Shimajiri et al. 2019; Tokuda et al. 2019). We emphasize that these high line-mass filaments are naturally generated in high shock velocity models over a relatively short time.

In Fig. 2.12, the filament line-mass functions are shown for the low shock velocity cases. The left panels illustrate the line-mass functions at approximately $t \simeq t_{\rm fil}$, while the right panels showcase them at the time when the maximum line-mass surpasses 100 $M_\odot$ pc^{-1}. In the results obtained from model v3GyTn, panel (i) indicates that the filaments have nearly critical line masses at $t = t_{\rm fil}$, consistent with our prior findings that attribute filament formation in this model to the type G mechanism. For model v3GyTy shown in panels (k) and (l), we find that the type C mode has the potential to produce super-critical filaments. However, the formation of these substantial filaments, around 100 $M_\odot$ pc^{-1}, demands a relatively extended duration,

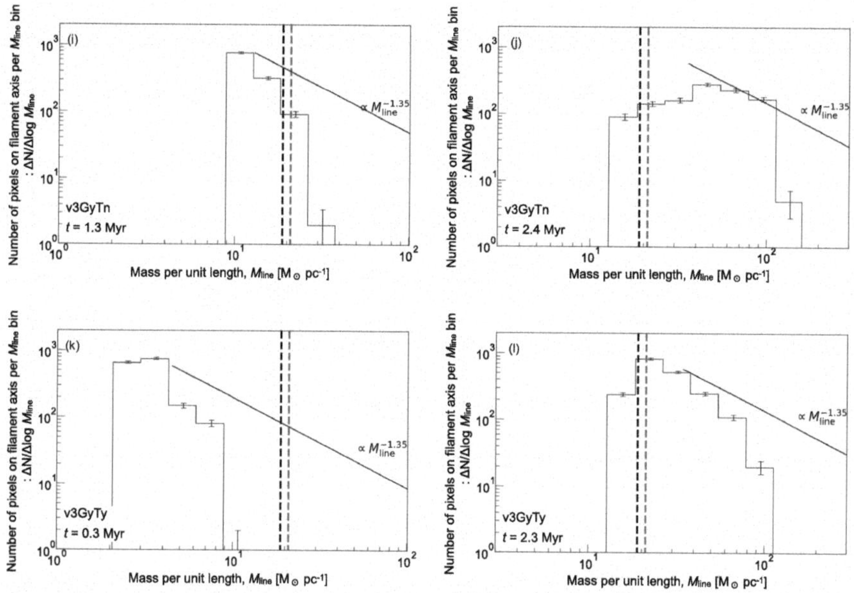

Fig. 2.12 Filament mass functions for model v3GyTn (*top panels* (i) and (j)) and v3GyTy (*bottom panels* (k) and (l)). *Left panels* (i) and (k): Filament mass functions at time $t = t_{\rm fil}$. *Right panels* (j) and (l): Results at time $t >$ 2 Myr. The lines' colors are the same as those defined in Fig. 2.11. Reproduced from Abe et al. (2021).

requiring more than 2 Myr following $t = t_{\rm fil}$. This emphasizes the time-intensive nature of generating such massive filaments in this model.

In Fig. 2.13, a comparison between the line-mass functions in models lacking self-gravity and those incorporating self-gravity is included. Panels (m) to (p) portray line-mass functions similar to those in panels (e) to (h), respectively. This comparison suggests that star-forming filaments can be generated irrespective of the presence of self-gravity. However, in panel (t), low line-mass filaments diminish, and a substantial portion of the filament mass function is truncated compared to panel (l). This indicates that low line-mass filaments, created by the type C mechanism, are transient. Moreover, it implies that self-gravity plays a pivotal role in the formation of massive filaments. Note that a column density threshold of $1.5 \times \bar{N}_{\rm sh}$ is applied for filament identification. Specifically, for model v3GnTy at t = 2.3 Myr, a threshold column density of 6.5×10^{21} cm^{-2} is imposed. Consequently, the identified filaments have a minimum line mass of 13 M$_\odot$ pc^{-1}, assuming a filament width of 0.1 pc.

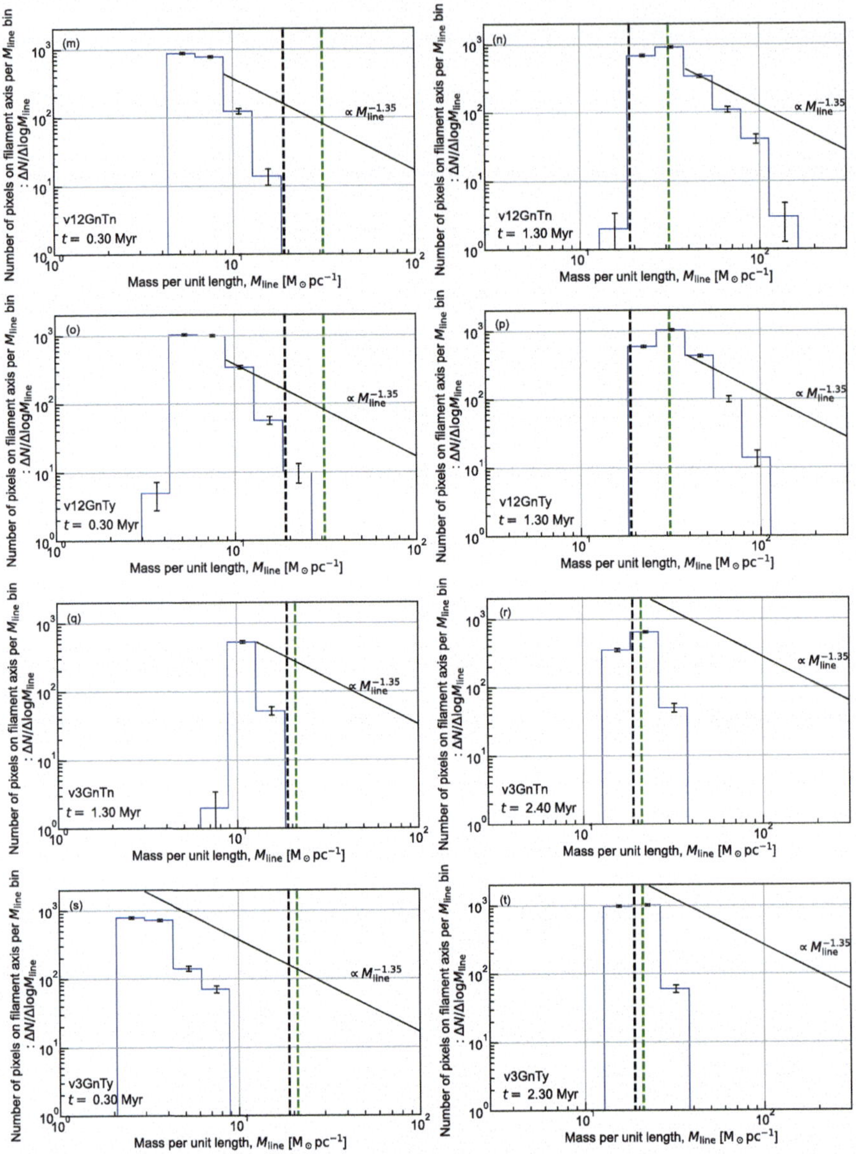

Fig. 2.13 Filament mass functions for no-self-gravitating models v12GnTn (panels m and n), v12GnTy (panels o and p), v12GnTn (panels q and r), and v12GnTy (panels s and t). To compare the histograms in the models with and without self-gravity, we show the histograms that are taken at the same times in Figs. 2.11 and 2.12. The lines' colors are the same as those defined in Fig. 2.11. Reproduced from Abe et al. (2021). © 2021. The Author(s). Published by the American Astronomical Society. CC BY 4.0

2.3.5 *Role of Shear: Angle Between Filaments and Magnetic Field*

In the type S mechanism, shear motions within turbulence contribute to the formation of faint filaments aligned parallel to the magnetic field. This concept has been studied by Soler et al. (2013), Soler and Hennebelle (2017), Körtgen and Soler (2020), and Inoue and Inutsuka (2016). To investigate the influence of shear flow, we compute probability distribution histograms based on the angles between the filaments and the mass-weighted average magnetic field within the projection plane. Given that we have already identified the filament skeletons, this computation of the angle is straightforward, allowing us to analyze the alignment between the filaments and the magnetic field orientation.

We present angle histograms for model v12GyTy at three different time points: 0.2 (panels a and b), 0.4 (panels c and d), and 0.8 (panels e and f) Myr, illustrated in Fig. 2.14. The top and bottom panels correspond to instances when $t \simeq t_{\rm fil}$ (refer to Sect. 2.3.3 and Fig. 2.8 panel (b) and $t \simeq t_{\rm ff}$ (Eq. 2.5), respectively. Panels (a), (c), and (e) depict angle histograms for filaments within the column density range of 0.5 $\bar{N}$sh to 1.5 $\bar{N}$sh, while panels (b), (d), and (f) illustrate histograms for filaments with $N > 1.5\ \bar{N}_{\rm sh}$. We note that a significant proportion of dense filaments in the right panels exhibit a perpendicular orientation to the magnetic field. Over time, the faint filaments depicted in the left panels transition from a perpendicular to a parallel alignment with the local magnetic field. This evolution can be naturally explained by the influence of turbulent shear (type S mode).

Figure 2.15 shows angle histograms for model v3GyTy at three-time points: 0.4 (panels g and h), 1.15 (panels i and j), and 2.0 (panels k and l) Myr. The top and bottom panels correspond to instances around $t_{\rm fil}$ and $t_{\rm ff}$, respectively. Similar to Fig. 2.14, the left panels represent filaments within the column density range of 0.5 $\bar{N}_{\rm sh}$ to 1.5 $\bar{N}_{\rm sh}$, while the right panels illustrate filaments exceeding 1.5 $\bar{N}_{\rm sh}$. The trend observed is similar to the higher shock velocity simulation results (v12GyTy; Fig. 2.14), although with more dispersed angle distributions, particularly noticeable for fainter filaments in the left panels. This increased dispersion is attributed to the slower evolution timescale in model v3GyTy compared to v12GyTy. Consequently, the fainter filaments have more time to be affected and stretched by the turbulent shear flows.

The findings corroborate that turbulent shear flow consistently extends gas structures over time, generating a filamentary structure characterized by low density, particularly outside regions of high column density. Notably, these filaments align with the magnetic field lines and maintain line masses below the critical threshold (Soler et al. 2013; Soler and Hennebelle 2017; Körtgen and Soler 2020; Inoue and Inutsuka 2016).

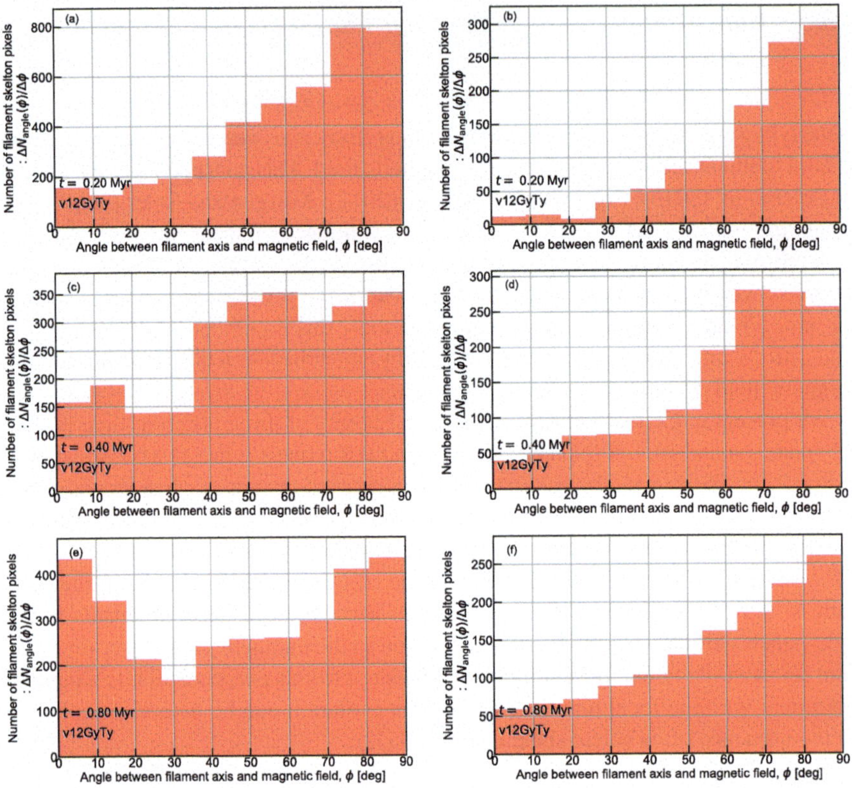

Fig. 2.14 Histogram of angles between filaments and magnetic field for model v12GyTy. From top to bottom, results at time $t = 0.2$, 0.4, and 0.8 Myr, respectively. *Top panels* (1) and (2) are results at filament formation time (see Sect. 2.3.3 and Fig. 2.8). *Bottom panels* (5) and (6) are results at close to free-fall time (Eq. 2.5). *Left panels* (1), (3), and (5): Results when we identify filaments in the column density range of $0.5\bar{N}_{\rm sh}$ to $1.5\bar{N}_{\rm sh}$. *Right panels* (2), (4), and (6): Results when the filament identification threshold column density is chosen to be $1.5\bar{N}_{\rm sh}$. Reproduced from Abe et al. (2021).

2.4 Discussion

2.4.1 PV Diagram in Type O Mechanism

To make it possible to compare the actual observations, we show the synthetic position-velocity (P-V) diagrams across the three major filaments in the results of the models v12GyTn (Fig. 2.16) and v12GyTy (Fig. 2.17). The color in the P-V diagram represents,

$$F(y, v) = \int \rho(x_0, y, z) \exp\left[-\frac{(v - v_z(x_0, y, z))^2}{c_s^2}\right] dz. \tag{2.7}$$

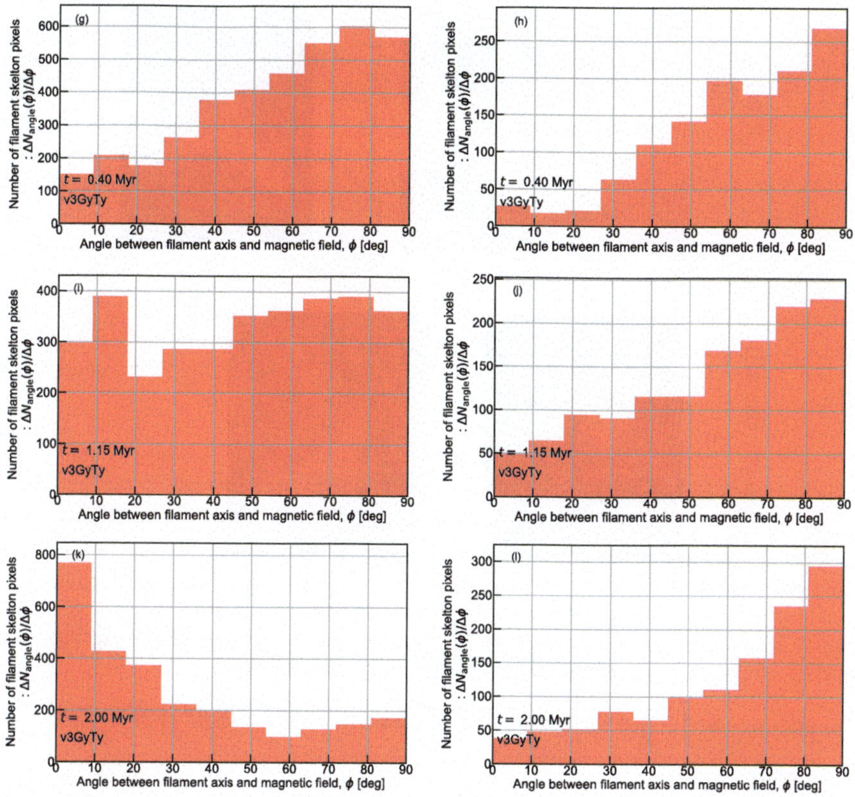

Fig. 2.15 Histogram of angles between filaments and magnetic field for model v3GyTy. From top to bottom, results at time $t = 0.4$, 1.15, and 2.0 Myr, respectively. *Top panels* (7) and (8) are results at filament formation time (see Sect. 2.3.3 and Fig. 2.9). *Bottom panels* (11) and (12) are results that exceed free-fall time (Eq. 2.5). *Left panels* (7), (9), and (11): Results when we identify filaments in the column density range of $0.5\bar{N}_{\rm sh}$ to $1.5\bar{N}_{\rm sh}$. *Right panels* (8), (10), and (12): Results when the filament identification threshold column density is chosen to be $1.5\bar{N}_{\rm sh}$. Reproduced from Abe et al. (2021).

The red regions in the PV diagrams correspond to filaments. In the Type O mechanism, it is shown that the gas flow behind curbed shock makes a V-shape like structure in P-V diagram (Inoue et al. 2018; Arzoumanian et al. 2018, in particular Sect. 4.2 and Fig. 12), while the turbulent mechanism does not (Chen et al. 2020). The V-shaped structure in major filaments of both v12GyTn and v12GyTy indicates that major filaments are created as a consequence of the oblique MHD shock compression mechanism when the gas collision velocity is high.

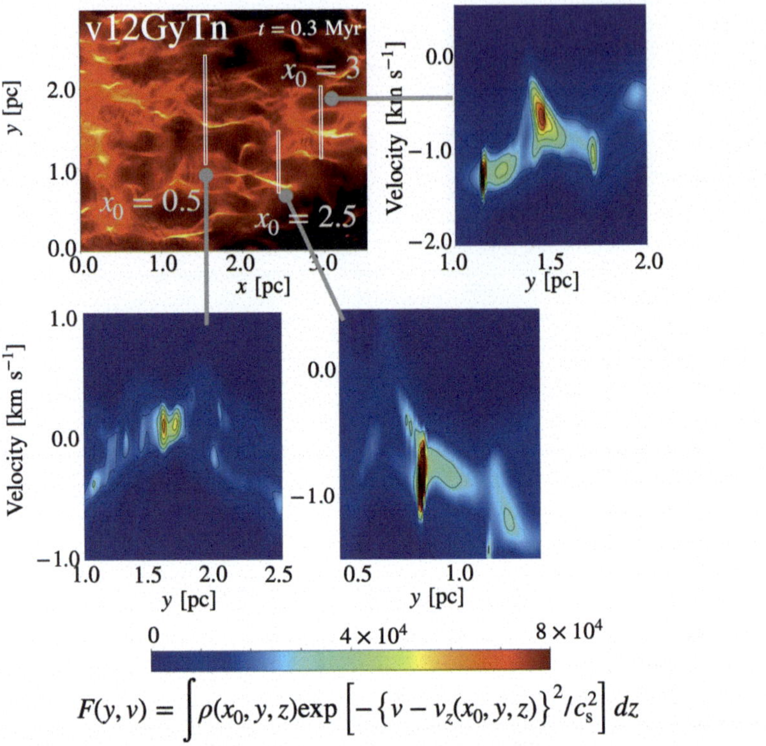

Fig. 2.16 Column density map and position-velocity (PV) diagrams of the result of the model v12GyTy at time $t = 0.3$ Myr. When we make PV diagrams, we specify x-position $x_0 = 0.5,\ 2.5,$ and 3.0 pc. The black line is the contour of $F(v, y)$ (see Eq. 2.7). The right bottom illustration is the schematic view of the filament formation process in the oblique MHD shock compression mechanism. The red regions in PV diagrams correspond to a filament. We can confirm the V-shape structure which is the supportive evidence for the oblique MHD shock compression mechanism (Arzoumanian et al. 2018)

2.4.2 *Estimation of Formation Time-Scales*

In this section, our objective is to estimate the growth timescale of filaments arising from both the type O (oblique shock effect) and type C (compressive flows from initial turbulence) mechanisms. This exploration aims to elucidate the underlying reasons for the shift in dominant filament formation mechanisms depending on shock velocity. It is essential to distinguish between the distinct characteristics and timescales of type O and type C mechanisms. We initiate our estimation by considering the filament formation timescale for the type O mechanism. As illustrated in panel (A) of Fig. 2.18, the oblique MHD shock model posits that the mass flux of post-shock gas flow to the filament is determined by $\rho_1 v_0 \sin\theta$, where θ represents the oblique

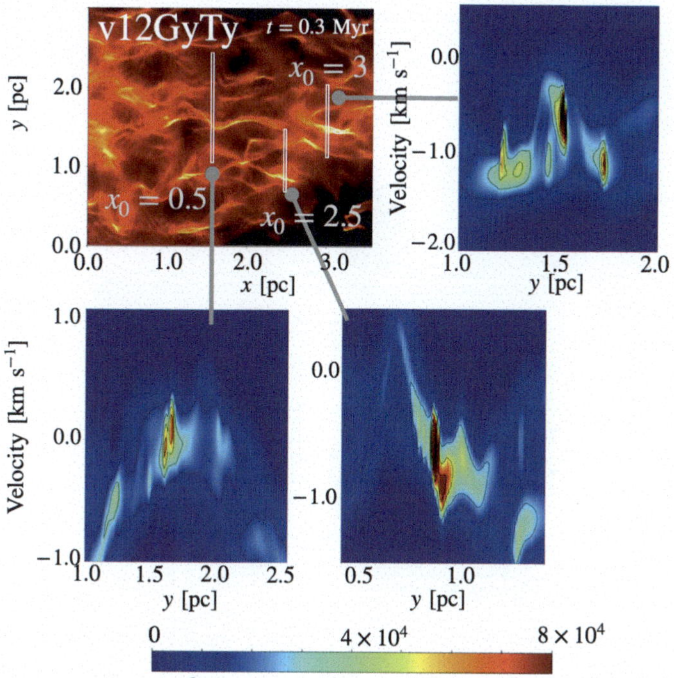

$$F(y, v) = \int \rho(x_0, y, z) \exp\left[-\left\{v - v_z(x_0, y, z)\right\}^2 / c_s^2\right] dz$$

Fig. 2.17 Column density map and position-velocity (PV) diagrams of the result of the model v12GyTy at time $t = 0.3$ Myr. The dense V-shaped structures also appear in this model, indicating that the oblique MHD shock compression mechanism occurs regardless of the initial turbulence

shock angle, leveraging the nearly conserved nature of the velocity component perpendicular to the shock normal across the shock.[3] If we assume constant width w of the filament, the line-mass of the filament after time t can be written as

$$M_{\rm line} = 2\rho_1 v_0 w t \sin\theta. \tag{2.8}$$

By substituting the isothermal strong MHD shock jump condition $\rho_1 \simeq \sqrt{2}\mathcal{M}_{\rm A}\rho_0$ (Eq. 2.1), the timescale $t_{\rm O}$, in which type O mechanism creates filament of the line-mass $M_{\rm line}$, is estimated in the following equation.

[3] The conservation of the parallel velocity component holds as an exact condition only under the assumption of no magnetic field. However, in the scenario involving fast shocks, the conservation of parallel velocity serves as a highly accurate approximation.

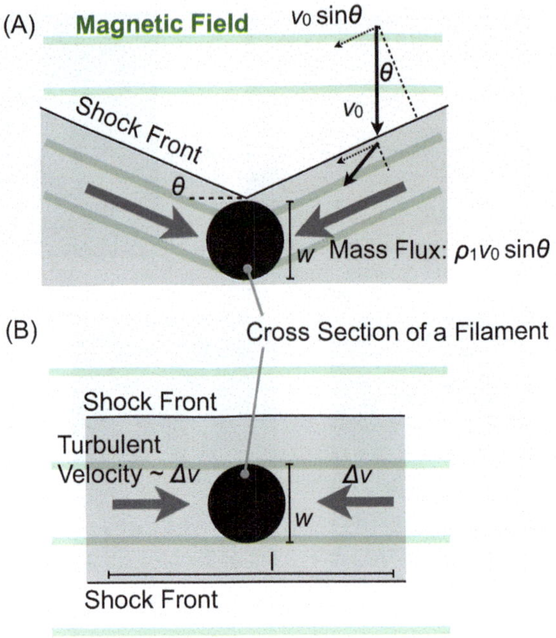

Fig. 2.18 Illustrations of the models used to estimate the filament formation timescale. The gray and black regions represent the post-shock layer and filament, respectively. (A): Schematic of the oblique MHD shock compression mechanism (type O), where w is the filament width, and θ is the oblique shock angle. The angle depends on the detail of the interaction between the shock and the gas clump that evolves into the filament and is roughly $\theta \sim 30°$ as indicated in Figs. 2.4 and 2.5. (B): Schematic of type C (compressive flows involved in initially given turbulence) mechanism, where l is the scale of the turbulent compressive flow related to the velocity of the flow Δv via Larson's law. Here we assume the width of the compressive flows as $w = 0.1$ pc. Reproduced from Abe et al. (2021).

$$
\begin{aligned}
t_{\mathrm{O}} &= \frac{M_{\mathrm{line}}}{2\rho_1 v_0 \sin\theta w} \\
&= \frac{M_{\mathrm{line}} B_0}{4\sqrt{2\pi}\rho_0^{3/2} v_0^2 w \sin\theta} \\
&= 0.3\ \mathrm{Myr}\ (\sin\theta)^{-1} \left(\frac{v_{\mathrm{sh}}}{7\ \mathrm{km\ s^{-1}}}\right)^{-2} \\
&\times \left(\frac{M_{\mathrm{line}}}{M_{\mathrm{line,cr}}}\right)\left(\frac{n_0}{100\ \mathrm{cm^{-3}}}\right)^{-3/2}\left(\frac{B_0}{10\ \mu\mathrm{G}}\right)\left(\frac{w}{0.1\ \mathrm{pc}}\right)^{-1},
\end{aligned}
\tag{2.9}
$$

where we have used the fact that $v_0 = v_{\mathrm{sh}}$.

In the context of type C filament formation depicted in panel (B) of Fig. 2.18, the formation occurs as a result of converging turbulent flows within the shock-compressed slab. Here, the velocity of these converging flows, represented as Δv, their scale denoted by l, and the width of the flow marked as h, are parameters. Given the origin of these converging flows in turbulence, Larson's law provides a relevant relationship: $\Delta v \sim V_{08}\,(l/L_1)^{0.5}$, where specific values such as $V_{08} = 0.8$ km s^{-1} and $L_1 = 1$ pc are established (Larson 1981). Then, the mass accumulation timescale is given by

$$t_C = l/\Delta v \sim l^{1/2} L_1^{1/2} / V_{08}. \tag{2.10}$$

By using l and w, we can estimate the line mass of the filament as

$$M_{\rm line} = \rho_1 l w. \tag{2.11}$$

By combining, Eqs. (2.10)–(2.11), the timescale t_C, in which the turbulent flow creates filament of the line-mass $M_{\rm line}$, can be estimated as

$$\begin{aligned} t_C &= \frac{M_{\rm line}}{\rho_1 w \Delta v} \\ &= \left(\frac{M_{\rm line} L_1 B_0}{2\sqrt{2\pi} w V_{08}^2 \rho_0^{3/2} v_{\rm sh}} \right)^{1/2} \\ &= 0.78\ {\rm Myr} \left(\frac{v_{\rm sh}}{7\ {\rm km\ s^{-1}}} \right)^{-1/2} \left(\frac{M_{\rm line}}{M_{\rm line,cr}} \right)^{1/2} \\ &\times \left(\frac{w}{0.1\ {\rm pc}} \right)^{-1/2} \left(\frac{\bar{n}_0}{100\ {\rm cm^{-3}}} \right)^{-3/4} \left(\frac{B_0}{10\ \mu{\rm G}} \right)^{1/2}. \end{aligned} \tag{2.12}$$

In the upper panel of Fig. 2.19, both timescales, t_O and t_C, are represented concerning the critical line-mass filament ($M_{\rm line} = M_{\rm line,cr}$) against the shock velocity. The lower panel illustrates the timescales for a filament with half the critical line mass ($M_{\rm line} = M_{\rm line,cr}/2$). Blue lines denote t_O assuming $\theta = 30°$, red lines represent t_C, and dashed lines indicate the free-fall time within the post-shock layer (Eq. 2.5). When the shock velocity is high ($v_{\rm sh} \gtrsim 5$ km s^{-1}, i.e., $v_{\rm coll} \gtrsim 8$ km s^{-1}), the relationship observed is $t_O < t_C < t_{\rm ff}$. Conversely, for lower shock velocities ($v_{\rm sh} \lesssim 5$ km s^{-1}, i.e., $v_{\rm coll} \lesssim 8$ km s^{-1}), the order becomes $t_C < t_{\rm ff} < t_O$. These findings corroborate well with the simulation outcomes, indicating the prevalence of the type O mechanism for higher shock velocities, while the type C mechanism dominates in scenarios with lower shock velocities.

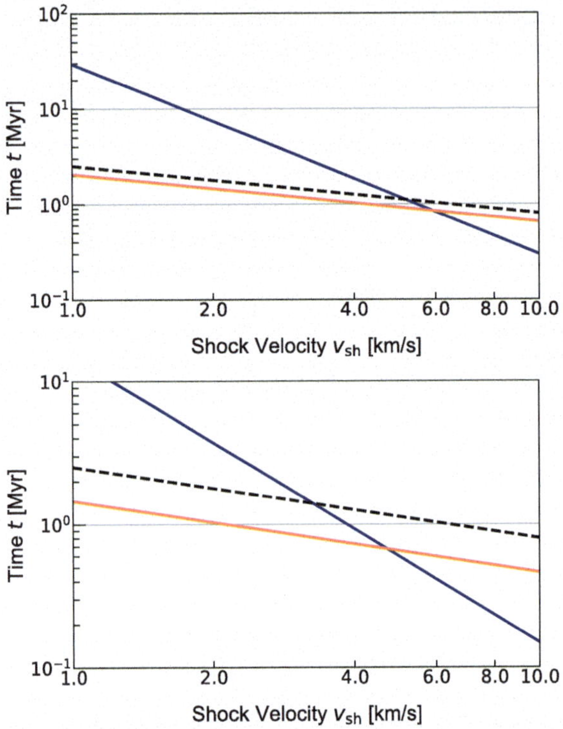

Fig. 2.19 Filament formation timescales as functions of shock velocity. The blue line shows the timescale for the type O mechanism given by Eq. (2.9), and the red line shows the type C timescale estimated by Eq. (2.12). The dashed line represents the free-fall time in the shocked region t_{ff} (Eq. 2.5). *Top panel*: Timescales required to reach the thermal critical line mass (e.g., Stodólkiewicz 1963; Ostriker 1964), i.e., $M_{line}(t) = M_{line,cr}$. *Bottom panel*: Timescales required to reach half the thermal critical line mass, i.e., $M_{line}(t) = M_{line,cr}/2$. Reproduced from Abe et al. (2021). © 2021. The Author(s). Published by the American Astronomical Society. CC BY 4.0

2.5 Summary

We conducted isothermal MHD simulations to investigate filament formation triggered by shock compression within a molecular cloud. Our findings reveal that for fast shocks ($v_{sh} \simeq 7$ km s^{-1}), the predominant mechanism driving the formation of star-forming filaments is the effect of oblique MHD shock-induced flows (type O), irrespective of initial turbulence or self-gravity. In cases of slower shocks ($v_{sh} \simeq 2.5$ km s^{-1}), transient filament formation is observed due to compressive flows associated with supersonic turbulence (type C). However, these resulting filaments tend to disperse unless the line-masses reach or exceed the thermal critical line-mass. Simulations with an initial input of strong turbulence, showing velocity dispersion exceeding ~5 km s^{-1}, do display local shock waves. While the type O process can theoretically occur in such scenarios, it doesn't align with what is conventionally defined as “turbulent” filamentation. In contrast, the type O mechanism typically emerges from a single compression event of a molecular cloud, commonly seen in interactions with large-scale, relatively fast shock waves like those from expanding HII regions, supernova remnants, or super-shells. Notably, observational differentiation between type O and type C mechanisms is feasible through characteristic features in position-velocity maps or curved magnetic fields, as highlighted in studies by (see, Arzoumanian et al. 2018; Tahani et al. 2018, 2019; Chen et al. 2020; Kandori et al. 2020a, b). In instances of low shock velocities and the absence of initial turbulence, filament formation (type G) primarily occurs via self-gravity-induced fragmentation of shock-compressed sheets over the gravitational fragmentation timescale of the dense sheet. Our simulations did not distinctly demonstrate filament formation through turbulent sheet-sheet collision (type I), as this mode appears to activate only in scenarios featuring weakly magnetized, uniform-density molecular clouds at the outset.

Our analytical models reveal that for shock velocities $\upsilon_{\rm sh} \gtrsim 5$ km s^{-1}, the dominant mechanism for filament formation is type O, while type C becomes more prominent for $\upsilon_{\rm sh} \lesssim 5$ km s^{-1}. This signifies a shift in the prevailing mode of filament formation based on the strength of the incident shock wave. Furthermore, our analysis of filament line-mass distributions illustrates that strong shock waves can naturally yield high-line-mass filaments similar to those observed in massive star-forming regions ($M_{\rm line} \gtrsim 100$ M$_\odot$ pc^{-1}) within a remarkably short time. Notably, such high line-mass filaments are inherently formed in high shock velocity models during the creation of dense compressed sheet-like regions.

References

Abe D, Inoue T, Inutsuka S-I, Matsumoto T (2021) ApJ 916:83. https://doi.org/10.3847/1538-4357/ac07a1, https://ui.adsabs.harvard.edu/abs/2021ApJ...916...83A

André P, Arzoumanian D, Könyves V, Shimajiri Y, Palmeirim P (2019) Astron Astrophys 629:1. https://doi.org/10.1051/0004-6361/201935915

Arzoumanian D et al (2011) A&A 529:L6. https://doi.org/10.1051/0004-6361/201116596, https://ui.adsabs.harvard.edu/abs/2011A&A...529L...6A

Arzoumanian D, Shimajiri Y, Inutsuka S, Inoue T, Tachihara K (2018) PASJ 70:96. https://doi.org/10.1093/pasj/psy095, https://ui.adsabs.harvard.edu/abs/2018PASJ...70...96A

Beresnyak A, Lazarian A, Cho J (2005) ApJ 624:L93. https://doi.org/10.1086/430702, https://ui.adsabs.harvard.edu/abs/2005ApJ...624L..93B

Bonne L et al (2020) A&A 644:A27. https://doi.org/10.1051/0004-6361/202038281, https://ui.adsabs.harvard.edu/abs/2020A&A...644A..27B

Chen C-Y, Mundy LG, Ostriker EC, Storm S, Dhabal A (2020) MNRAS 494:3675. https://doi.org/10.1093/mnras/staa960, https://ui.adsabs.harvard.edu/abs/2020MNRAS.494.3675C

Crutcher RM (2012) ARA&A 50:29. https://doi.org/10.1146/annurev-astro-081811-125514, https://ui.adsabs.harvard.edu/abs/2012ARA&A..50...29C

Dedner A, Kemm F, Kröner D, Munz CD, Schnitzer T, Wesenberg M (2002) JCoPh 175:645. https://doi.org/10.1006/jcph.2001.6961

Elmegreen BG, Scalo J (2004) ARA&A 42:211. https://doi.org/10.1146/annurev.astro.41.011802.094859, https://ui.adsabs.harvard.edu/abs/2004ARA&A..42..211E

Fukui Y et al (2019) ApJ 886:14. https://doi.org/10.3847/1538-4357/ab4900, https://ui.adsabs.harvard.edu/abs/2019ApJ...886...14F
Fukui Y, Habe A, Inoue T, Enokiya R, Tachihara K (2020) https://ui.adsabs.harvard.edu/abs/2020arXiv200905077F, arXiv:2009.05077
Heiles C, Crutcher R (2005) Cosm Fields, Magn 137–182. https://doi.org/10.1007/11369875_7
Heyer MH, Brunt CM (2004) ApJ 615:L45. https://doi.org/10.1086/425978, https://ui.adsabs.harvard.edu/abs/2004ApJ...615L..45H
Inoue T, Fukui Y (2013) ApJ 774:L31. https://doi.org/10.1088/2041-8205/774/2/L31, https://ui.adsabs.harvard.edu/abs/2013ApJ...774L..31I
Inoue T, Hennebelle P, Fukui Y, Matsumoto T, Iwasaki K, Inutsuka S-I (2018) PASJ 70:S53. https://doi.org/10.1093/pasj/psx089, https://ui.adsabs.harvard.edu/abs/2018PASJ...70S..53I
Inoue T, Inutsuka S (2016) ApJ 833:10. https://doi.org/10.3847/0004-637X/833/1/10, https://ui.adsabs.harvard.edu/abs/2016ApJ...833...10I
Inutsuka S (2001) ApJ 559:L149. https://doi.org/10.1086/323786, https://ui.adsabs.harvard.edu/abs/2001ApJ...559L.149I
Kandori R et al (2020a) ApJ 892:128. https://doi.org/10.3847/1538-4357/ab7b68, https://ui.adsabs.harvard.edu/abs/2020ApJ...892..128K
Kandori R et al (2020b) ApJ 900:20. https://doi.org/10.3847/1538-4357/abaab3, https://ui.adsabs.harvard.edu/abs/2020ApJ...900...20K
Koch EW, Rosolowsky EW (2015) MNRAS 452:3435. https://doi.org/10.1093/mnras/stv1521, http://adsabs.harvard.edu/abs/2015MNRAS.452.3435K
Körtgen B, Soler JD (2020) MNRAS 499:4785. https://doi.org/10.1093/mnras/staa3078, https://ui.adsabs.harvard.edu/abs/2020MNRAS.499.4785K
Larson RB (1981) MNRAS 194:809. https://doi.org/10.1093/mnras/194.4.809, https://ui.adsabs.harvard.edu/abs/1981MNRAS.194..809L
Matsumoto T (2007) PASJ 59:905. https://doi.org/10.1093/pasj/59.5.905, https://ui.adsabs.harvard.edu/abs/2007PASJ...59..905M
Matsumoto T, Dobashi K, Shimoikura T (2015) ApJ 801:77. https://doi.org/10.1088/0004-637X/801/2/77, https://ui.adsabs.harvard.edu/abs/2015ApJ...801...77M
Miyoshi T, Kusano K (2005) JCoPh 208:315. https://doi.org/10.1016/j.jcp.2005.02.017
Nagai T, Inutsuka S-I, Miyama SM (1998) ApJ 506:306. https://doi.org/10.1086/306249, https://ui.adsabs.harvard.edu/abs/1998ApJ...506..306N
Ostriker J (1964) ApJ 140:1056. https://doi.org/10.1086/148005, https://ui.adsabs.harvard.edu/abs/1964ApJ...140.1056O
Scalo J, Elmegreen BG (2004) ARA&A 42:275. https://doi.org/10.1146/annurev.astro.42.120403.143327, https://ui.adsabs.harvard.edu/abs/2004ARA&A..42..275S
Shimajiri Y, André P, Ntormousi E, Men'shchikov A, Arzoumanian D, Palmeirim P (2019) A&A 632:A83. https://doi.org/10.1051/0004-6361/201935689, https://ui.adsabs.harvard.edu/abs/2019A&A...632A..83S
Soler JD, Hennebelle P (2017) A&A 607:A2. https://doi.org/10.1051/0004-6361/201731049, https://ui.adsabs.harvard.edu/abs/2017A&A...607A...2S
Soler JD, Hennebelle P, Martin PG, Miville-Deschênes MA, Netterfield CB, Fissel LM (2013) ApJ 774:128. https://doi.org/10.1088/0004-637X/774/2/128, https://ui.adsabs.harvard.edu/abs/2013ApJ...774..128S
Stodólkiewicz JS (1963) Acta Astron 13:30. https://ui.adsabs.harvard.edu/abs/1963AcA....13...30S
Tahani M, Plume R, Brown JC, Kainulainen J (2018) A&A 614:A100. https://doi.org/10.1051/0004-6361/201732219, https://ui.adsabs.harvard.edu/abs/2018A&A...614A.100T
Tahani M, Plume R, Brown JC, Soler JD, Kainulainen J (2019) A&A 632:A68. https://doi.org/10.1051/0004-6361/201936280, https://ui.adsabs.harvard.edu/abs/2019A&A...632A..68T
Tokuda K et al (2019) ApJ 886:15. https://doi.org/10.3847/1538-4357/ab48ff, https://ui.adsabs.harvard.edu/abs/2019ApJ...886...15T
Tomisaka K (2014) ApJ 785. https://doi.org/10.1088/0004-637X/785/1/24

Chapter 3
The Effect of Shock Wave Duration on Star Formation and the Initial Condition of Massive Cluster Formation

3.1 Short Introduction

As massive stars and clusters hold a crucial role in the evolution and dynamics of galaxies, it is significant to understand their formation mechanism. It has been observed that stars tend to emerge within dense molecular filaments, irrespective of their mass. Extensive research into these formations reveals a noteworthy detail: the predominant mechanism appears to vary depending on the velocity of the shock-triggering filament formation (Abe et al. 2021). Among these mechanisms, 'Type O' stands out, requiring high Mach number shocks to produce substantial filaments within the compressed layers (Abe et al. 2021).

Comparative studies between numerical simulations of shock compression in molecular clouds from cloud collisions and observational data have highlighted the potential of strong MHD shock compression in generating massive cores (Inoue and Fukui 2013; Inoue et al. 2018; Sakre et al. 2021). However, these investigations primarily assumed prolonged shock waves, akin to collisions between sizable clouds. In reality, the duration of shock waves varies considerably based on the specific circumstances. In contrast, simulations involving unmagnetized cloud collisions have indicated that the formation of gravitationally bound cores might not occur with short-duration shock waves (Takahira et al. 2014). Thus, a comprehensive exploration to understand the influence of shock duration on the formation dynamics of filaments and cores appears imperative.

Enokiya et al. (2021) demonstrated a correlation between the peak column density of a star-forming region and the count of OB stars within the system. Notably, massive star clusters featuring over 10 OB-type stars tend to associate with clouds having a peak column density surpassing 10^{23} cm^{-2}. Understanding the underlying physical reasons behind this specific threshold column density of 10^{23} cm^{-2} is imperative. It is worth noting the relationship between the mean column density $\bar{N}_{H_2}$ and the shock duration t_{dur}, as $\bar{N}_{H_2} = \bar{n}L$ and $t_{dur} = L/v_{sh}$ yield $\bar{N}_{H_2} \propto t_{dur}$, where $\bar{n}$, L, and v_{sh} denote the mean number density of a cloud, cloud size, and shock

D. Abe, *Formation and Evolution of Star-Forming Filaments in Molecular Clouds*,
Springer Theses, https://doi.org/10.1007/978-981-97-6414-3_3

velocity, respectively. Hence, it's plausible to consider that shock duration serves as a crucial parameter influencing the determination of the peak column density. Previous simulations employing high column density cores/clumps as initial conditions have successfully shown the formation of massive stars/clusters (Bonnell et al. 2004; Krumholz and McKee 2008; Krumholz et al. 2009, 2012). These simulations indicate that the presence of a high column density initial core/clump is pivotal for the formation of massive stars/clusters. For instance, in Krumholz et al. (2009), a dense core of 100 $M_{\odot}$ within 0.1 pc was used as an initial condition, a value justified by observations of the protostellar object IRAS 05358+3543. However, such extremely dense cores are not commonly found in observations. Recent molecular investigations of IRAS 05358+3543 by Yamada et al. (2022) have revealed evidence of two colliding molecular clouds at scales spanning several parsecs. These collisions lead to the formation of a dense filament encompassing the core IRAS 05358+3543. This study strongly supports the notion that the formation of high-mass stars necessitates robust external triggers, such as the collision of molecular clouds. It is essential to note that the duration of such collisions should exceed the free-fall time calculated from the mean density in the shock-compressed layer to generate the requisite high column density for subsequent star formation. This underscores the critical role of both external triggers and their duration in the formation of high-mass stars.

In this chapter, we investigate the correlation between shock duration and subsequent star formation by regulating the gas influx into the numerical domain. Our aim is to understand the impact of shock duration on the genesis of massive clusters and to elucidate the threshold peak column density for their formation. Earlier studies on this subject have yielded conflicting conclusions. Inoue et al. (2018) and Abe et al. (2021) showcased the formation of massive cores/stars through fast MHD shock compression by assuming prolonged shock duration, albeit with assumptions of either long or infinite shock wave duration. Conversely, studies by Takahira et al. (2014) and Sakre et al. (2022), focusing on collisions between relatively smaller clouds, suggested that fast shock compression prevents the formation of gravitationally bound cores. To resolve these discrepancies, we examine how shock duration influences the formation of massive cores/stars. The structure of this paper unfolds as follows: Sect. 3.2 details the setup of our simulations, while Sect. 3.3 encompasses the demonstration and interpretation of our findings. We delve into discussing the peak column density within the shock-compressed layer through a developed theoretical model in Sect. 3.4. Finally, a comprehensive summary of our results is presented in Sect. 3.5.

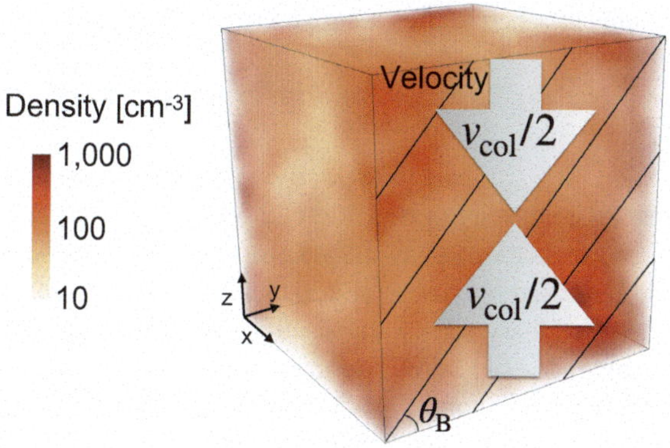

Fig. 3.1 Schematic of the initial condition. The color bar represents the density magnitude, the black lines represent the initial magnetic field lines, the θ_B represents the angle between the y-axis and the initial magnetic field lines, and the gray arrows represent the orientations of the converging flows. Reproduced from Abe et al. (2021).

3.2 Setup for Simulations

We perform three-dimensional isothermal[1] MHD simulations including self-gravity using SFUMATO code (Matsumoto 2007). Our setup is the same as that of Abe et al. (2021) but includes a finite duration for shock compression. Using a cubic domain of $L_{box} = 6$ pc, we fill it with nonuniform gas characterized by a mean density n_0 and fluctuations reflecting a power spectrum of $(\log \rho)\mathrm{k}^2 \propto k^{-4}$ due to supersonic turbulence (Larson 1981; Beresnyak et al. 2005; Elmegreen and Scalo 2004; Scalo and Elmegreen 2004; Heyer and Brunt 2004). Gas temperature is fixed at 10 K ($c_s \sim 0.2$ km/s). The numerical domain is divided into uniform 512^3 cells, providing a spatial resolution of $\Delta x = 1.2 \times 10^{-2}$ pc resulting from a domain size of 6 pc. The initial velocity field, $\mathbf{v}(x, y, z)$, is determined as $\mathbf{v}_{turb}(x, y, z) - (v_{col}/2) \tanh[(z - L_{box}/2)/0.1]\hat{\mathbf{z}}$, where, $\mathbf{v}_{turb}$ represents turbulent velocity with a dispersion of 1.0 km s^{-1} and a power spectrum following Larson's law (Larson 1981), $v_k^2 \propto k^{-4}$. This setup generates a gas collision at the plane $z = L_{box}/2$. The chosen initial magnetic field strength is $B_0 = 10$ μG (Heiles & Crutcher 2005; Crutcher 2012), with an inclination angle θ_B to the y-z plane set at either 45° or 60° (Fig. 3.1).

[1] The isothermal approach is effective for depicting compression but might not adequately represent the subsequent expansion regions post t_{dur}. In these regions, the density could decrease to as low as 1 cm^{-3}, causing the gas to transition into a warm atomic state. Such a transition may lead to effects like evaporation flow due to gas heating and thermal conduction. However, given the expansion occurs at sonic speed due to pressure reduction, the influence of non-isothermal effects may not be significant.

The velocity fields at z-boundaries are given by

$$\mathbf{v}^{\pm}_{\text{boundary}}(x, y) = \left[\pm v_{\text{col}}/2\hat{\mathbf{z}} + \mathbf{v}_{\text{turb}}(x, y, z)\right] \exp[-(t - t_{\text{stop}})/0.1\text{Myr}], \quad (3.1)$$

where "$\mathbf{v}^{+}_{\text{boundary}}$" and "$\mathbf{v}^{-}_{\text{boundary}}$" are the velocity field at $z = 0$, and 6 pc, respectively, and "v_{col}" is the relative velocity of two flows colliding at $z = 3$ pc. These boundary conditions enforce the cessation of gas inflows around $t = t_{\text{stop}}$, differing significantly from the methodology used by Abe et al. (2021). The density and turbulence fields incoming from the boundary match the spatial distribution present in the initial condition. For the magnetic field, free boundary conditions are imposed at $z = 0$ and $z = L_{\text{box}}$. Periodic boundary conditions are employed for all physical variables at $x = 0$, $x = L_{\text{box}}$, $y = 0$, and $y = L_{\text{box}}$ boundary planes.

We conducted simulations for a total of 18 models, each uniquely named starting with "v" denoting the collision velocity: "3," "5," "14," "18", and "24" (in km/s), followed by the shock duration designated as "t" spanning from 0.23 to 1.9 Myr. Additionally, models varying in initial magnetic field angle and mean density are distinguished by the tags "ang60" and "d300," representing $\theta_{\text{B}} = 60°$ and $\bar{n}_0 = 300\,\text{cm}^{-3}$, respectively. Refer to Table 3.1 for the detailed parameter sets used in our simulations.

Table 3.1 Model parameters

Model name	v_{coll} [km s^{-1}]	v_{sh} [km s^{-1}]	t_{stop} [Myr]	t_{dur} [Myr]	θ_{B}	$\bar{n}_0$ [cm^{-3}]	Δx [pc]
v3t1.4	3	2.2	0.0	1.4	45°	100	1.2×10^{-2}
v5t0.94	5	3.2	0.0	0.94	45°	100	1.2×10^{-2}
v5t1.5	5	3.2	0.7	1.5	45°	100	1.2×10^{-2}
v14t0.39	14	7.7	0.0	0.39	45°	100	1.2×10^{-2}
v14t0.40ang60	14	7.5	0.0	0.40	60°	100	1.2×10^{-2}
v14t0.66	14	7.7	0.3	0.66	45°	100	1.2×10^{-2}
v14t0.84	14	7.7	0.5	0.84	45°	100	1.2×10^{-2}
v14t1.0	14	7.7	0.7	1.0	45°	100	1.2×10^{-2}
v14t1.1d300	14	7.4	0.7	1.1	45°	300	1.2×10^{-2}
v14t1.5	14	7.7	1.2	1.5	45°	100	1.2×10^{-2}
v14t1.7ang60	14	7.5	1.4	1.7	60°	100	1.2×10^{-2}
v14t1.9	14	7.7	1.7	1.9	45°	100	1.2×10^{-2}
v18t0.68	18	9.7	0.4	0.68	45°	100	1.2×10^{-2}
v24t0.23	24	13	0.0	0.23	45°	100	1.2×10^{-2}
v24t0.43	24	13	0.2	0.43	45°	100	1.2×10^{-2}
v24t0.61	24	13	0.4	0.61	45°	100	1.2×10^{-2}
v24t1.6	24	13	1.4	1.6	45°	100	1.2×10^{-2}
v24t1.6AMR	24	13	1.4	1.6	45°	100	5.9×10^{-3}
v24t1.8	24	13	1.7	1.8	45°	100	1.2×10^{-2}

In regions prone to gravitational collapse, we introduce sink particles (Matsumoto et al. 2015; Inoue et al. 2018). The criteria for generating sink particles depend on the resolution, with a threshold density for sink creation set at $5.6\times10^4\,\mathrm{cm}^{-3}$ in the current 512^3 cell simulations, a lower threshold than in previous studies. To investigate potential resolution and sink creation density effects, we conducted AMR simulations. Following the Jeans criterion for refinement (Truelove et al. 1997), we set $\Delta x \leq \lambda_\mathrm{J}/8$, where $\lambda_\mathrm{J} = \pi^{1/2} c_\mathrm{s}/\sqrt{G\rho}$ represents the Jeans length, resulting in a minimum cell size of $\Delta x = 6$ pc/1024 $= 5.9 \times 10^{-3}$ pc. The threshold density for sink particle formation in the AMR run is $2.2\times10^5\,\mathrm{cm}^{-3}$. Notably, the results are almost the same as those in the non-AMR runs, as detailed in Sect. 3.

3.3 Results

3.3.1 Column Density Maps

3.3.1.1 Short Duration Case

Figure 3.2 shows snapshots of the column density map for model v24t0.23 at intervals of $t = 0.10, 0.20, 0.30$, and 0.60 Myr, depicted in panels (a)–(d) and (a')–(d') for y-z and x-y planes, respectively. Panels (a')–(d') illustrate the development of dense filaments over time. Abe et al. (2021) categorized filament formation mechanisms into distinct types. According to their classification, the filamentation observed in model v24t0.23 is type O, indicative of the oblique MHD shock compression mechanism. Type O filamentation is known to manifest behind robust shocks with $v_\mathrm{sh} \gtrsim 5\,\mathrm{km\ s^{-1}}$, while the shock waves generated in this model possess v_sh approximately around $13\,\mathrm{km\ s^{-1}}$. In panel (a) of Fig. 3.3, the blue, red, and green markers correspond to gas masses within regions possessing densities greater than 10^3, 10^4, and $10^5\,\mathrm{cm}^{-3}$, respectively. Panels (b')–(d') in Fig. 3.2 and panel (a) in Fig. 3.3 collectively illustrate a decline in the mass of dense gas over a 0.2–0.3 Myr duration. This evolution arises from the expansion of the shock-compressed layer, evident in Fig. 3.2d, leading to a reduction in pressure and diffusion of dense filaments. The time at which the compression layer starts to expand t_dur can be written using t_stop as

$$t_\mathrm{dur} \simeq t_\mathrm{stop} + \frac{L_\mathrm{box}/2 - v_1 t_\mathrm{dur}}{v_\mathrm{col}/2} \tag{3.2}$$

$$\simeq \frac{t_\mathrm{stop} + L_\mathrm{box}/v_\mathrm{col}}{1+\sqrt{2}v_\mathrm{alf,0}/v_\mathrm{col}}. \tag{3.3}$$

where $v_1 \simeq \bar{v}_\mathrm{Alf,\perp}/\sqrt{2}$ is the shock velocity at the rest frame of the compression layer, and $\bar{v}_\mathrm{Alf,\perp} = B_{0,\perp}/\sqrt{4\pi\bar{\rho}_0}$ is the mean Alfvén velocity. $B_{0,\perp} = B_0\cos(\theta_\mathrm{B})$ is the initial magnetic field strength perpendicular to the shock normal. The second term on the right-hand side in Eq. (3.2) describes the delay in time for the gas to reach

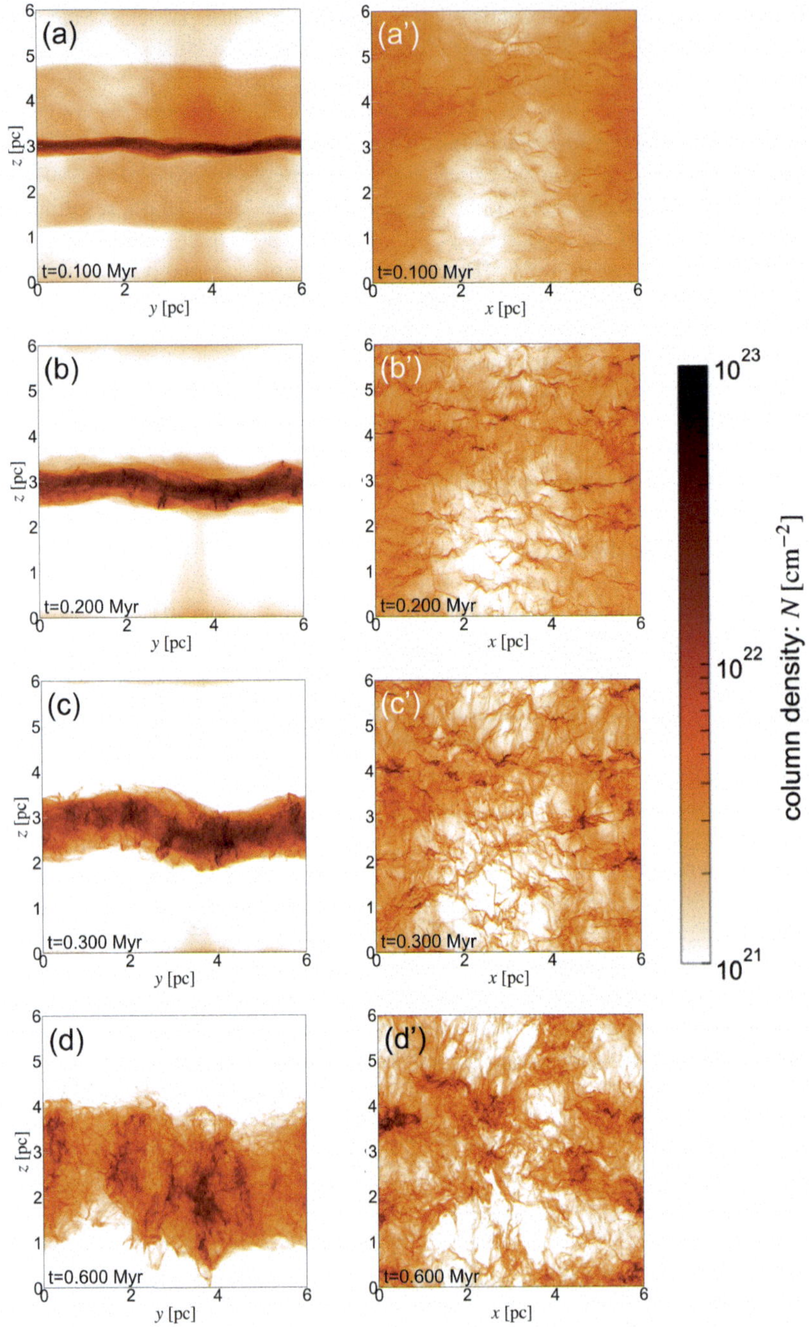

Fig. 3.2 Column density maps of the result of model v24t0.23 at time $t = 0.3$, 0.8, 1.6, and 1.8 Myr (from top to bottom). *Left row* (panels a, b, c, and d): y-z plane column densities. *Right row* (panels a', b', c', and d'): x-y plane column densities. Reproduced from Abe et al. (2021).

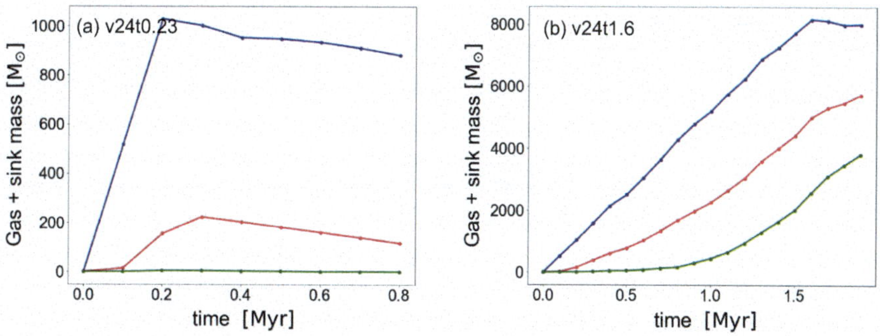

Fig. 3.3 Panel a: the evolution of the total dense gas mass in the model v24t0.23. The blue, red, and green points represent the total masses with densities above 10^3, 10^4, and 10^5 cm^{-3}, respectively. Panel b: the same as panel (a), but for model v24t1.6. Reproduced from Abe et al. (2021).

the compression layer after $t_{\rm stop}$. Specifically, for model v24t0.23, $t_{\rm dur} \simeq 0.23$ Myr corresponds to the time when the dense gas mass in Fig. 3.3 reaches its maximum.

The free-fall time in the postshock layer, which gives the timescale for self-gravitating sheet fragmentation (Nagai et al. 1998), can be estimated as

$$
\begin{aligned}
t_{\rm ff} &= \sqrt{\frac{1}{2\pi G \bar{\rho}_1}} \\
&= \sqrt{\frac{\bar{v}_{\rm Alf,\perp}}{2\sqrt{2}\pi G \bar{\rho}_0 \bar{v}_{\rm sh}}} \\
&= \sqrt{\frac{B_{0,\perp}}{4\sqrt{2}\pi^{3/2} G \bar{\rho}_0^{3/2}\left(v_{\rm coll}/2 + B_0/\sqrt{8\pi\bar{\rho}_0}\right)}} \\
&\simeq 1.0\ {\rm Myr} \left(\frac{B_{0,\perp}}{10\ \mu{\rm G}}\right)^{1/2} \left(\frac{\bar{n}_0}{100\ {\rm cm}^{-3}}\right)^{-3/4} \\
&\times \left[\left(\frac{v_{\rm coll}}{12\ {\rm km/s}}\right) + 0.17\left(\frac{B_{0,\perp}}{10\ \mu{\rm G}}\right)\left(\frac{\bar{n}_0}{100\ {\rm cm}^{-3}}\right)^{-1/2}\right]^{-1/2},
\end{aligned}
\tag{3.4}
$$

where $\bar{\rho}_1 \simeq \sqrt{2}\mathcal{M}_{\rm A}\bar{\rho}_0$ is the mean density of the shocked layer (e.g., Inoue and Fukui 2013), and $\bar{v}_{\rm sh} = v_{\rm coll}/2 + v_1 \simeq v_{\rm coll}/2 + B_0/\sqrt{8\pi\bar{\rho}_0}$ represents the mean shock velocity. Given that $t_{\rm dur} < t_{\rm ff} \simeq 0.68$ Myr in model v24t0.23, the shock compression layer expands before gravitational fragmentation occurs, leading to the absence of sink particle formation.

3.3.1.2 Long Duration Case

Similar to Figs. 3.2 and 3.4 shows snapshots of the column density map for model v24t1.6 at times t = 0.30, 0.80, 1.60, and 1.80 Myr. As the shock velocity remains consistent with the previous model (v24t0.23), dense filamentary structures form via the type O mechanism. The "+" symbols in Fig. 3.4 indicate the positions of the sink particles. Panel (b) of Fig. 3.3 illustrates the evolution of the dense gas mass in the numerical domain, where the blue, red, and green points represent the masses of regions with densities exceeding 10^3, 10^4, and 10^5 cm^{-3}, respectively, as in panel (a). It should be noted that the vertical axis range differs from that of panel (a). In this model as well, the shock-compressed layer expands after $t = t_{\rm dur} \simeq 1.6$ Myr, as in model v24t0.23. However, the compression duration exceeds the free-fall-time calculated using the mean density in the layer $t_{\rm ff} \simeq 0.58$ Myr (Eq. 2.5). This extended compression time enables the filaments to coalesce due to gravitational contraction within the compression layer. Consequently, unlike model v24t0.23, the dense gas mass continues to increase even after $t_{\rm dur}$ (panel a in Fig. 3.3).

3.3.2 *Peak Column Density Versus Dense Gas Mass*

Observations indicate a correlation between the peak column density of star-forming clouds and the number of OB stars in the system (Enokiya et al. 2021). In this context, we present a comparative analysis between the results of our simulations and the observed data. We consider two types of mass in the numerical domain as proxies for star formation: the total mass of sink particles, denoted as $M_{\rm sink,tot}(t)$, and the total gas mass in the region with $A_V > 8$ mag ($N_{\rm H_2} \simeq 7.8 \times 10^{21}$ cm^{-2}), denoted as $M_{\rm Av8}(t)$ (Lada et al. 2010). The sum of these masses, $M_{\rm dense}(t) \equiv M_{\rm sink,tot}(t) + M_{\rm Av8}(t)$, is proportional to the total mass of stars in the system. Figure 3.5 illustrates the evolution of $M_{\rm dense}(t)$, with panels (a) and (b) presenting the results for models v24t0.23 and v24t1.6, respectively. In our simulations, $M_{\rm dense}(t)$ saturates at late times, allowing us to define a mass associated with star formation. We fit $M_{\rm dense}(t)$ using a trial function $\alpha \tanh[2\pi(t - t_0)/t_{\rm w}] + \beta$, as depicted by the red lines in Fig. 3.5. This fit enables us to determine the time $t_{\rm sat} \equiv t_0 + t_{\rm w}/4$ (dashed lines in Fig. 3.5), representing the total dense gas mass that characterizes the star formation activity in the system. Here, α, t_0, $t_{\rm w}$, β are fitting parameters. It's noteworthy that $t_{\rm sat}$ corresponds to the time when the fitting curve reaches approximately $\alpha \tanh[2\pi(t_{\rm sat} - t_0)/t_{\rm w}] + \beta = \tanh(\pi/2) + \beta \simeq 0.92\alpha + \beta$. In other words, this value indicates the mass corresponding to $(0.92 + 1.0) \times 100/2 = 96\%$ of the increment in the trial function.

In general, the peak column density is significantly influenced by the spatial grid (or, equivalently, spatial resolution) employed in the derivation process. Consequently, the observed peak column densities of candidates for cloud-cloud collision (CCC), as illustrated in Fig. 9 of Enokiya et al. (2021), may potentially possess independent spatial resolutions. Despite this, a discernible correlation between the peak

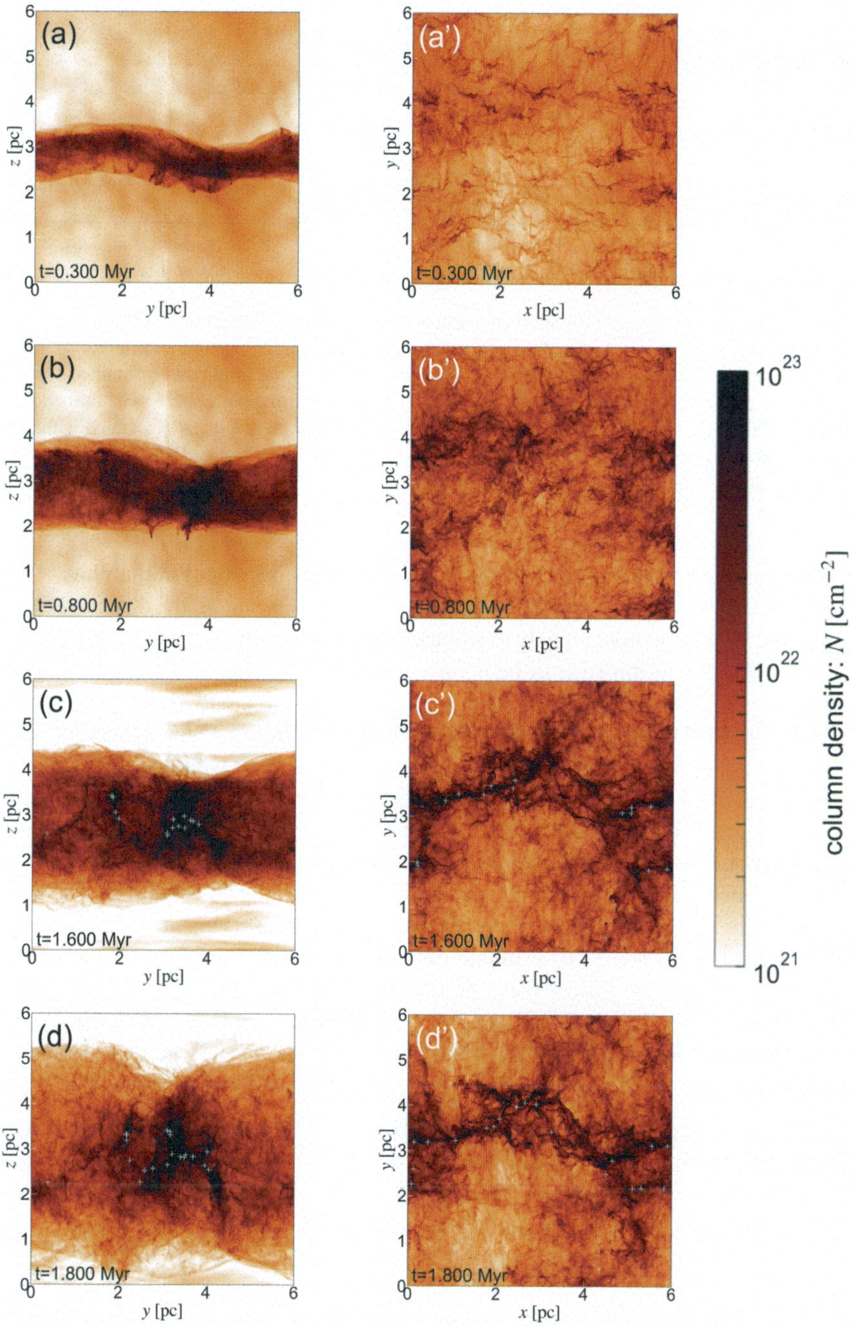

Fig. 3.4 Column density maps in the result of model v24t1.6 at time $t = 0.3, 0.8, 1.6$, and 1.8 Myr (from top to bottom). *Left row* (panels a, b, c, and d): y-z plane column densities. *Right row* (panels a', b', c', and d'): x-y plane column densities. Reproduced from Abe et al. (2021).

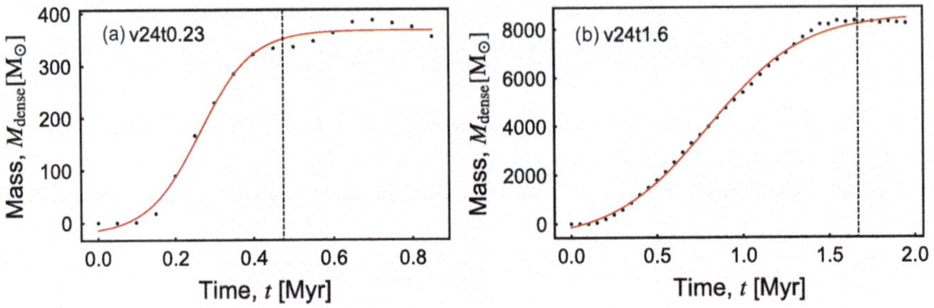

Fig. 3.5 Evolution of $M_{\rm dense}(t)$, which is the sum of the total mass of sink particles ($M_{\rm sink,tot}(t)$) and the mass of gas with column density larger than 7.8×10^{21} cm^{-2} ($M_{\rm Av8}(t)$) for the simulation results of models v24t0.23 (left) and v24t1.6 (right). A red line represents the fitting curve obtained using a trial function $\alpha \tanh[2\pi(t - t_0)/t_{\rm w}] + \beta$, where α, t_0, $t_{\rm w}$, β are fitting parameters. The dashed line denotes the time $t_{\rm sat}$, which is defined as the time at which the total gas mass associated with star formation is measured. Reproduced from Abe et al. (2021). © 2021. The Author(s). Published by the American Astronomical Society. CC BY 4.0

column density and the number of O- and B-type stars is evident in the plots. This implies that there is likely uniformity in spatial resolutions among the observed CCC candidates. We have meticulously examined the spatial resolution among these CCC candidates and determined it to be approximately $\sim$ the size of the smaller cloud divided by 10. Assuming a size of 5 pc for the smaller cloud (refer to Fig. 9 in Fukui et al. 2021), we estimate the typical spatial resolution of the observed CCC candidates to be 0.5 pc. Hence, prior to computing the peak column density of the system, we apply a smoothing technique to the column density structure using a Gaussian kernel function with a width of 0.5 pc. In panel (a) of Fig. 3.6, we present a scatter plot illustrating the relationship between the peak column density and the mass of dense gas $M_{\rm dense}(t_{\rm sat})$ for all simulations outlined in Table 3.1. The coloration of the data points corresponds to the ratio $t_{\rm dur}/t_{\rm ff}$, serving as an indicator of the impact of self-gravity. To estimate the number of OB stars formed within the numerical domain, we follow these procedures: First, we multiply $M_{\rm dense}(t_{\rm sat})$ by the star formation efficiency SFE$_{\rm dense}$ $\sim$0.1, as suggested by Fukui et al. (2021), to obtain the total mass of stars. Next, the fraction of OB-type stars in the total stellar mass, denoted as $f_{\rm OB} = \left(\int_{10M_\odot}^{150M_\odot} dM\, M F_{\rm Kroupa}\right) / \left(\int_{0.01M_\odot}^{150M_\odot} dM\, M F_{\rm Kroupa}\right) = 0.19$, is applied to derive the total mass of OB stars expected in the system. It is essential to note that we consider stars with masses exceeding the early-B type, i.e., those larger than approximately 10 $M_\odot$. Here, $F_{\rm Kroupa} \equiv dN/dM$ is an IMF proposed by Kroupa (2001). Therefore, the estimation of the number of OB stars is

$$N_{\rm OB} \simeq M_{\rm dense}(t_{\rm sat}) \times {\rm SFE}_{\rm dense} \times f_{\rm OB} \times \left(\frac{\int_{10M_\odot}^{150M_\odot} dM\, M F_{\rm Kroupa}}{\int_{10M_\odot}^{150M_\odot} dM\, F_{\rm Kroupa}} \right)^{-1}, \quad (3.5)$$

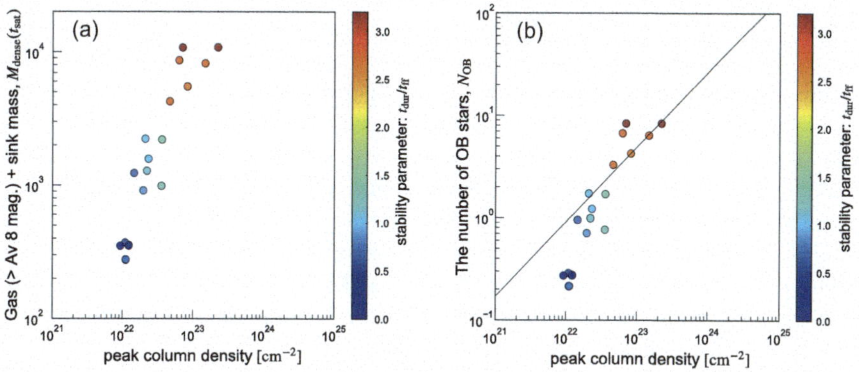

Fig. 3.6 Panel a: Scatter plot of the peak column density versus the dense gas mass $M_{\rm dense}(t_{\rm sat})$ for all runs summarized in Table 2.1. Before computing the peak column density of the system, we take a smoothing of the column density using the Gaussian kernel function of a width of 0.5 pc, corresponding to the typical beam width for massive star-forming regions. The color of the points represents $t_{\rm dur}/t_{\rm ff}$ indicating the influence of self-gravity. Panel b: Scatter plot of the peak column density versus the estimated number of OB-type stars $N_{\rm OB}$. $N_{\rm OB}$ is estimated using Eq. (3.5). The black line is the fitting line in Fig. 9 (b) of Enokiya et al. (2021). Reproduced from Abe et al. (2021).

where $\int_{10M_\odot}^{150M_\odot} dM\, M F_{\rm Kroupa} / \int_{10M_\odot}^{150M_\odot} dM\, F_{\rm Kroupa}$ on the right-hand side is the typical mass of OB stars. Panel (b) of Fig. 3.6 is derived from panel (a) through the utilization of Eq. (3.5). The outcome reveals a correlation akin to that observed by Enokiya et al. (2021), despite our simulations encompassing a more restricted range of peak column density compared to the observations. The power-law index obtained by fitting the data in panel (b) is 0.75, exhibiting a noteworthy agreement with the observational value of 0.73±0.11 (Enokiya et al. 2021). The black line represents the fitting line from Fig. 9 (b) of Enokiya et al. (2021). Observationally, it is well-established that the formation of massive star clusters occurs in clouds with a peak column density surpassing $N_{\rm peak} \geq 10^{23}$ cm^{-2} (Fukui et al. 2016). Our simulations agree with the identified threshold peak column density (as evident in the region with $N_{\rm peak} \sim 10^{23}$cm^{-2} in panel (b) of Fig. 3.6). The stability parameter, defined as the ratio of the shock duration to the free-fall time of the compressed layer $t_{\rm dur}/t_{\rm ff}$, indicates that the formation of massive stars is anticipated in scenarios where the dense shock-compressed layer persists over the free-fall time. We find that when $N_{\rm peak}$ exceeds 10^{23} cm^{-2}, the stability parameter $\gtrsim 2$. Remarkably, the identified threshold peak column density bears a resemblance to the proposed threshold column density for fragmentation suppression, which is 1 g cm^{-2}, as suggested by Krumholz and McKee (2008).

3.3.3 Sink Mass Histogram

Figure 3.7 shows the histograms of sink particle masses at $t = t_{\rm sat}$ for the models v14t0.38, v14t0.84, and v14t1.9. In our simulations, sink particles correspond to gravitationally collapsing cores. It is important to note that Fig. 3.7 does not represent a core mass function due to the limited resolution of low-mass cores in our simulations. Model v14t1.9, characterized by an extended flow duration of $t_{\rm dur}/t_{\rm ff} \simeq 2.6$, exhibits the formation of a massive sink particle with a mass of 355 $M_\odot$ and the potential for the creation of OB-type stars. The mass and number of sink particles increase with $t_{\rm dur}$ as expected, but it is noteworthy that only the long-duration model v14t1.9, with $t_{\rm dur}/t_{\rm ff} \gtrsim 2$, demonstrates active and substantial massive star formation.

Figure 3.8 illustrates a sink mass histogram at $t = t_{\rm sat}$ for the models v24t1.6 and v24t1.6AMR, comparing the outcomes with and without Adaptive Mesh Refinement (AMR). The count of sink particles with masses around 1 $M_\odot$ has increased due to the AMR calculation's ability to capture self-gravitational fragmentation of short-wavelength modes in dense filaments. However, it is noteworthy that the total masses of massive sink particles (those exceeding $10 M_\odot$) are nearly identical between v24t1.6 ($M_{\rm sink} = 1803\, M_\odot$) and v24t1.6AMR ($M_{\rm sink} = 2048\, M_\odot$). This suggests that the results presented in Sect. 3.2 remain largely unchanged, even with the improvement in spatial resolution facilitated by AMR.

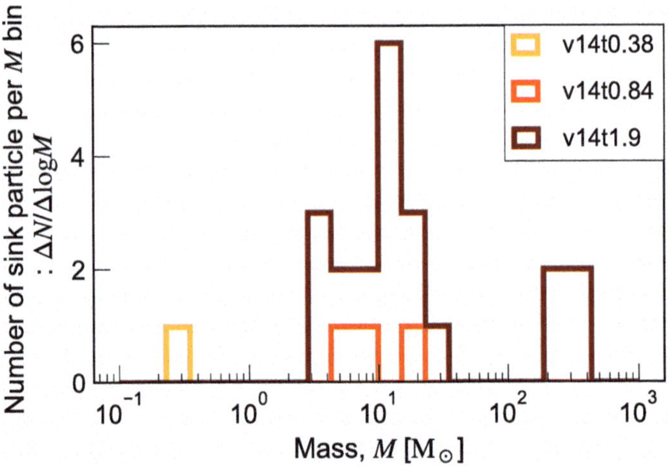

Fig. 3.7 Sink mass histogram at $t = t_{\rm sat}$ in the results of models v14t0.38 (orange), v14t0.84 (red), and v141.9 (brown). Reproduced from Abe et al. (2021).

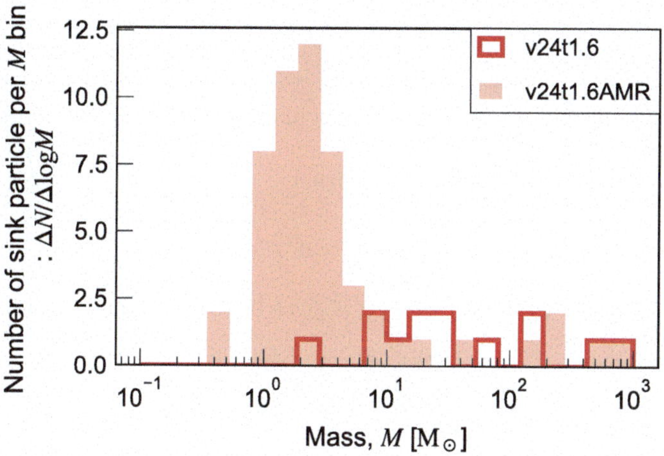

Fig. 3.8 Sink mass histogram at $t = t_{sat}$ in the models v24t1.6 (red) and v24t1.6AMR. (filled-in region). Reproduced from Abe et al. (2021).

3.4 Discussion

3.4.1 Estimation of Peak Column Density

In Sect. 3.3.2, we asserted that self-gravitational collapse should attain a peak column density of 10^{23} cm^{-2}. In the context of observational column density maps consisting of approximately ~500 × 500 pixels, this peak column density roughly corresponds to being 5σ away from the mean of the Probability Density Function (PDF) obtained from observations. In this analysis, we estimate the peak column density by employing a theory of the gas density PDF in a turbulent medium. This theory indicates that gravitational collapse within shock-compressed layers is necessary to achieve a peak column density of 10^{23} cm^{-2}. The PDF of the gas density in a turbulent medium shows log-normal PDF of the form (Passot and Vázquez-Semadeni 2003; Padoan et al. 2014)

$$p_s(s) = \frac{1}{\sqrt{2\pi\sigma_s^2}} \exp\left(-\frac{(s - s_1)^2}{2\sigma_s^2}\right). \tag{3.6}$$

where $s \equiv \ln(\rho/\rho_1)$, ρ_1 and s_1 represent the mean density and mean logarithmic density in the shocked layer, respectively. The latter is related to the standard deviation σ_s as $s_1 = -\sigma_s^2/2$. Molina et al. (2012) find that the standard deviation σ_s is determined using the turbulence Mach number $\mathcal{M}$, plasma beta β, and the ratio of solenoidal mode to compressive mode b (pure solenoidal forcing gives $b = 1/3$ and pure compressive forcing gives $b = 1$) as

$$\sigma_s^2 = \ln\left(1 + b^2\mathcal{M}^2\frac{\beta}{\beta+1}\right). \tag{3.7}$$

Here, we estimate the maximum possible column density of a cloud formed by the gas collision. The density PDF can have maximum width if we substitute the collision velocity for the turbulence velocity dispersion

$$\mathcal{M}_{\max} \equiv v_{\rm col}/(2c_{\rm s}). \tag{3.8}$$

The beta of the shock-compressed layer is given by the shock jump condition of the isothermal MHD as

$$\beta_{\rm sh} = \frac{8\pi\rho_1 c_s^2}{B_1^2} \simeq \frac{8\pi\rho_0 c_s^2}{\sqrt{2}\mathcal{M}_{\rm A,max} B_0^2}, \tag{3.9}$$

where $\rho_1 \simeq \sqrt{2}\mathcal{M}_{\rm A,max}\rho_0$, $B_1 \simeq \sqrt{2}\mathcal{M}_{\rm A,max}B_0$, and $\mathcal{M}_{\rm A,max} = v_{\rm col}/(2v_{\rm alf,0})$ are the density, magnetic field in the compression layer, and the Alfvén Mach number of the shock. If we adopt the parameter $b = 0.4$, as anticipated in the shock-compressed layer (Kobayashi et al. 2022), we can determine $\sigma_{\rm s,max}$ and the density Probability Density Function (PDF).

Numerical experiments of supersonic isothermal turbulence and observations (Goodman et al. 2009) show that the PDF of column density is also close to log-normal (Federrath et al. 2010)

$$p(\eta) = \frac{1}{\sqrt{2\pi\sigma_\eta^2}}\exp\left(-\frac{(\eta-\eta_1)^2}{2\sigma^2}\right), \tag{3.10}$$

where $\eta \equiv \ln(N/N_1)$. N and N_1 are column density and the mean column density, respectively. By comparing s_1 and η_1 in numerical simulations by Federrath et al. (2010), η_1 is approximately expressed as $\eta_1 \sim 0.5s_1$. The standard deviation is also approximated by $\sigma \sim 0.5\sigma_{\rm s}$. These relationships can be used to convert density PDF to column-density PDF.

The peak column density is defined as 5σ away from the mean

$$N_{5œ} \equiv \exp(\eta_1 + 5\sigma_{,\max})N_1, \tag{3.11}$$

where $\sigma_{,\max} \equiv 0.5\sigma_{\max}$. Upon substituting the parameters for the strong collision case ($v_{\rm col} = 10$ km s^{-1}, $\bar{n}0 = 100$ cm^{-3}, and $B_0 = 10$ μG), the calculated peak column density $N_{5œ}$ is 8.7×10^{22} cm^{-2}, which falls below the threshold of 10^{23} cm^{-2}. This suggests that even if the velocity dispersion of turbulence in the shock-compressed layer is overestimated, the peak column density generated solely by turbulence is not expected to surpass 10^{23} cm^{-2}. Consequently, gravitational collapse of the shock layer is deemed necessary to achieve this value. In essence, the shock-compressed layer must be sustained until gravitational collapse is initiated, leading to the formation of a massive star cluster capable of elevating the peak column density.

3.4.2 Which Is Advantageous for Massive Star Formation, Fast or Slow Collisions?

The recent study by Sakre et al. (2022) suggests that there exists a maximum collision speed that can trigger star formation in a cloud collision scenario, which is consistent with findings from other studies. It is commonly observed that a high collision speed can result in a short shock duration. However, it is crucial to note that a high shock velocity is not necessarily detrimental to induced star formation. As we elaborate below, a high shock velocity generally has positive implications for star formation: The duration of the shock can be estimated by $t_{\rm dur} = L/v_{\rm sh}$, where L is the spatial scale and $v_{\rm sh}$ is the typical velocity of the flows. If we consider a supersonic turbulent gas flow collision as the origin of the shock, Larson's law indicates $v_{\rm sh} \propto L^{0.5}$ or $L \propto v_{\rm sh}^2$, and consequently, $t_{\rm dur} = L/v_{\rm sh} \propto v_{\rm sh}$. Given that the average free-fall time of the shocked layer is $t_{\rm ff} \propto v_{\rm sh}^{-1/2}$ as per Eq. (2.5), the ratio of the shock duration to the postshock free-fall time can be expressed as $t_{\rm dur}/t_{\rm ff} \propto v_{\rm sh}^{3/2}$. This implies that a faster shock leads to a longer duration in units of free-fall time, indicating that larger clouds naturally result in more active star formation. On the contrary, if we fix the scale of the flow L and use a different scaling $t_{\rm dur} = L/v_{\rm sh} \propto v_{\rm sh}^{-1}$, we obtain an opposite result $t_{\rm dur}/t_{\rm ff} \propto v_{\rm sh}^{-1/2}$. Thus, under the assumption of a fixed cloud scale, a faster shock is associated with a negative effect on star formation. It is essential to note that this is a consequence of artificially fixing L and does not represent a general trend. The overall trend suggests that, in a natural setting where cloud size can vary, a faster shock generally leads to more active star formation.

3.5 Summary

In our study, we conducted simulations of the shock compression of molecular clouds using SFUMATO, a 3D isothermal magnetohydrodynamic code that incorporates self-gravity. The focus of our investigation was to examine the impact of the shock-compressed layer's duration on filament and star formation by manipulating the shock duration as a controlled variable. We conducted a detailed analysis of the correlation between the peak column density and the estimated number of OB-type stars, aiming to gain insights into the initial conditions necessary for the formation of massive stars. To validate our findings, we compared our results with the observations presented by Enokiya et al. (2021). Our primary conclusions from this study are as follows.

1. The comparison between the short-duration and long-duration models reveals distinct outcomes. In the short-duration model, the shock leads to the formation of filaments, but this is subsequently followed by expansion and evaporation after the shock duration timescale. Conversely, the long-duration model results in the formation of massive filaments, fostering conditions conducive to star formation.

2. If the observed peak column density exceeds 10^{23} cm^{-2}, it suggests that the number of OB stars expected to form in the shock-compressed layer can reach the order of ten, indicating the potential for massive cluster formation. This observation aligns with the findings of Enokiya et al. (2021), particularly evident in the region with $N_{peak} \sim 10^{23} cm^{-2}$ in panel (b) of Fig. 3.6. According to a simple theoretical model, achieving such a high peak column density is possible only when the shock-compressed layer undergoes gravitational collapse.
3. The crucial requirement for activating massive star formation is that the duration of shock compression must be at least two free-fall times in the compressed layer ($t_{dur}/t_{ff} \gtrsim 2$). Importantly, this conclusion remains robust and is not significantly influenced by spatial resolution considerations.

References

Abe D, Inoue T, Inutsuka S-I, Matsumoto T (2021) ApJ 916:83. https://doi.org/10.3847/1538-4357/ac07a1, https://ui.adsabs.harvard.edu/abs/2021ApJ...916...83A

Beresnyak A, Lazarian A, Cho J (2005) ApJ 624:L93. https://doi.org/10.1086/430702, https://ui.adsabs.harvard.edu/abs/2005ApJ...624L..93B

Bonnell IA, Vine SG, Bate MR (2004) MNRAS 349:735. https://doi.org/10.1111/j.1365-2966.2004.07543.x, https://ui.adsabs.harvard.edu/abs/2004MNRAS.349..735B

Crutcher RM (2012) ARA&A 50:29. https://doi.org/10.1146/annurev-astro-081811-125514, https://ui.adsabs.harvard.edu/abs/2012ARA&A..50...29C

Elmegreen BG, Scalo J (2004) ARA&A 42:211. https://doi.org/10.1146/annurev.astro.41.011802.094859, https://ui.adsabs.harvard.edu/abs/2004ARA&A..42..211E

Enokiya R, Torii K, Fukui Y (2021) PASJ 73:S75. https://doi.org/10.1093/pasj/psz119, https://ui.adsabs.harvard.edu/abs/2021PASJ...73S.75E

Federrath C, Roman-Duval J, Klessen RS, Schmidt W, Mac Low MM (2010) A&A 512:A81. https://doi.org/10.1051/0004-6361/200912437, https://ui.adsabs.harvard.edu/abs/2010A&A...512A..81F

Fukui Y et al (2016) ApJ 820:26. https://doi.org/10.3847/0004-637X/820/1/26, https://ui.adsabs.harvard.edu/abs/2016ApJ...820...26F

Fukui Y, Habe A, Inoue T, Enokiya R, Tachihara K (2021) PASJ 73:S1. https://doi.org/10.1093/pasj/psaa103, https://ui.adsabs.harvard.edu/abs/2021PASJ...73S...1F

Goodman AA, Pineda JE, Schnee SL (2009) ApJ 292:91. https://doi.org/10.1088/0004-637X/692/1/91, https://ui.adsabs.harvard.edu/abs/2009ApJ...692...91G

Heiles C, Crutcher R (2005) Cosm. Fields, Magn. 137–182. https://doi.org/10.1007/11369875_7

Heyer MH, Brunt CM (2004) ApJ 615:L45. https://doi.org/10.1086/425978, https://ui.adsabs.harvard.edu/abs/2004ApJ...615L..45H

Inoue T, Fukui Y (2013) ApJ 774:L31. https://doi.org/10.1088/2041-8205/774/2/L31, https://ui.adsabs.harvard.edu/abs/2013ApJ...774L.31I

Inoue T, Hennebelle P, Fukui Y, Matsumoto T, Iwasaki K, Inutsuka S-I (2018) PASJ 70:S53. https://doi.org/10.1093/pasj/psx089, https://ui.adsabs.harvard.edu/abs/2018PASJ...70S.53I

Kobayashi MIN, Inoue T, Tomida K, Iwasaki K, Nakatsugawa H (2022). https://ui.adsabs.harvard.edu/abs/2022arXiv220300699K, arXiv:2203.00699

Kroupa P (2001) MNRAS 322:231. https://doi.org/10.1046/j.1365-8711.2001.04022.x, https://ui.adsabs.harvard.edu/abs/2001MNRAS.322..231K

Krumholz MR, Klein RI, McKee CF (2012) ApJ 754:71. https://doi.org/10.1088/0004-637X/754/1/71, https://ui.adsabs.harvard.edu/abs/2012ApJ...754...71K

Krumholz MR, Klein RI, McKee CF, Offner SSR, Cunningham AJ (2009) Science 323:754. https://doi.org/10.1126/science.1165857, https://ui.adsabs.harvard.edu/abs/2009Sci...323..754K
Krumholz MR, McKee CF (2008) Nature 451:1082. https://doi.org/10.1038/nature06620, https://ui.adsabs.harvard.edu/abs/2008Natur.451.1082K
Lada CJ, Lombardi M, Alves JF (2010) ApJ 724:687. https://doi.org/10.1088/0004-637X/724/1/687, https://ui.adsabs.harvard.edu/abs/2010ApJ...724..687L
Larson RB (1981) MNRAS 194:809. https://doi.org/10.1093/mnras/194.4.809, https://ui.adsabs.harvard.edu/abs/1981MNRAS.194..809L
Matsumoto T (2007) PASJ 59:905. https://doi.org/10.1093/pasj/59.5.905, https://ui.adsabs.harvard.edu/abs/2007PASJ...59..905M
Matsumoto T, Dobashi K, Shimoikura T (2015) ApJ 801:77. https://doi.org/10.1088/0004-637X/801/2/77, https://ui.adsabs.harvard.edu/abs/2015ApJ...801...77M
Molina FZ, Glover SCO, Federrath C, Klessen RS (2012) MNRAS 423:2680. https://doi.org/10.1111/j.1365-2966.2012.21075.x, https://ui.adsabs.harvard.edu/abs/2012MNRAS.423.2680M
Nagai T, Inutsuka S-I, Miyama SM (1998) ApJ 506:306. https://doi.org/10.1086/306249, https://ui.adsabs.harvard.edu/abs/1998ApJ...506..306N
Padoan P, Federrath C, Chabrier G, Evans NJI, Johnstone D, Jørgensen JK, McKee CF, Nordlund Å (2014) In: Beuther H, Klessen RS, Dullemond CP, Henning T (eds) Protostars and Planets VI, p 77
Passot T, Vázquez-Semadeni E (2003) A&A 398:845. https://doi.org/10.1051/0004-6361:20021665, https://ui.adsabs.harvard.edu/abs/2003A&A...398..845P
Sakre N, Habe A, Pettitt AR, Okamoto T (2021) PASJ 73:S385. https://doi.org/10.1093/pasj/psaa059, https://ui.adsabs.harvard.edu/abs/2021PASJ...73S.385S
Sakre N, Habe A, Pettitt AR, Okamoto T, Enokiya R, Fukui Y, Hosokawa T (2022). https://ui.adsabs.harvard.edu/abs/2022arXiv220507057S, arXiv:2205.07057
Scalo J, Elmegreen BG (2004) ARA&A 42:275. https://doi.org/10.1146/annurev.astro.42.120403.143327, https://ui.adsabs.harvard.edu/abs/2004ARA&A..42..275S
Takahira K, Tasker EJ, Habe A (2014) ApJ 792:63. https://doi.org/10.1088/0004-637X/792/1/63, https://ui.adsabs.harvard.edu/abs/2014ApJ...792...63T
Truelove JK, Klein RI, McKee CF, Holliman JHI, Howell LH, Greenough JA (1997) ApJ 489:L179. https://doi.org/10.1086/310975, https://ui.adsabs.harvard.edu/abs/1997ApJ...489L.179T
Yamada RI et al (2022). MNRAS. https://doi.org/10.1093/mnras/stac1087, https://ui.adsabs.harvard.edu/abs/2022MNRAS.tmp.1113Y

Chapter 4
Filament Evolution Process

4.1 Short Introduction

Type G, C, and O are the fundamental formation mechanisms of supercritical filaments, which are essential for star formation. The common feature of these mechanisms is that filaments are formed by gas flow along the local magnetic field line in a shocked-compressed sheet. Molecular emission-line observations provide evidence of the perpendicular accretion onto filaments (Palmeirim et al. 2013; Shimajiri et al. 2019; Chen et al. 2020). In particular, Shimajiri et al. (2019) reported the occurrence of accretion onto filaments in a shocked sheet. The accretion rate $\dot{M}_{\rm line}$ is estimated in the range from a few 10 to a few 100 $M_\odot$ Myr^{-1} pc^{-1} (Palmeirim et al. 2013; Bonne et al. 2020). The accretion timescale $t_{\rm acc} \equiv M_{\rm line}/\dot{M}_{\rm line}$ is comparable to self-gravitational fragmentation time-scale in filaments, implying that accretion is essential for the evolution of filaments.

The filament width is one of the significant quantities to determine the initial condition for star formation. The critical line mass considering magnetic support depends on the width of a filament (Tomisaka 2014). According to linear theory, the self-gravitational fragmentation length scale of the filament depends on the filament width (Stodólkiewicz 1963; Inutsuka & Miyama 1992a). Arzoumanian et al. (2018) revealed that the characteristic width of *Herschel* Gould Belt filaments is 0.1 pc (see also Koch and Rosolowsky 2015), and similar results have been reported by $C^{18}O$ line-emission observations (Orkisz et al. 2019; Suri et al. 2019). Remarkably, the filaments maintain their width, regardless of their line-mass exceeding 100 $M_\odot$ pc^{-1}. If we consider only thermal support against gravity, such a high-line-mass structure cannot maintain a length scale of 0.1 pc. Several authors studied the effects of turbulence and/or magnetic field and have shown that sub-critical and mildly supercritical filaments have a width of 0.1 pc (Fischera and Martin 2012; Auddy et al. 2016; Priestley and Whitworth 2022; Federrath 2016; Ntormousi and Hennebelle 2019), however, the reason for the constant width of filaments, especially for massive filaments, remains as a mystery. We need to understand the origin of the universal width,

D. Abe, *Formation and Evolution of Star-Forming Filaments in Molecular Clouds*,
Springer Theses, https://doi.org/10.1007/978-981-97-6414-3_4

especially for massive filaments, by examining the detailed process of gas accretion flows onto the filament.

Given that massive filaments are formed in the post-shock layer threaded by the strong magnetic field, the filament surface is naturally bound by the slow shocks. This unphysical feature stems from the ideal approximation and the resulting discontinuous treatment of the shock. To know the physical scale length of the SSI, we consider a non-ideal effect. Since the corrugation of the shock front generally produces turbulent flows behind the shock (e.g., Inoue et al. 2012; Inoue and Inutsuka 2012b), we can expect that the SSI will deposit additional energy to the filament.

In molecular clouds, ambipolar diffusion is effective and potentially modifies the SSI dynamics. The Reynolds number of the ambipolar diffusion is given as

$$\mathcal{R}_{\rm AD} = \frac{4\pi\gamma\rho_{\rm n}\rho_{\rm i}v\ell}{B^2}, \tag{4.1}$$

where, $\gamma_{\rm in} \equiv \langle\sigma_{\rm in}v_{\rm in}\rangle/(m+m_{\rm i}) = 3.5\times10^{13}\ {\rm cm^3\ g^{-1}\ s^{-1}}$ and $\rho_{\rm i}$ denotes the ion mass density. $\sigma_{\rm in}$, $v_{\rm in}$, m, and $m_{\rm i}$ represent the Langevin cross-section, the relative velocity between a neutral molecule and ion, mean molecule mass, and mean ion mass, respectively. Assuming a balance between the ionization by cosmic rays and the recombination, $\rho_{\rm i}$ can be expressed as $C\rho^{1/2}$. The characteristic length scale below which the effect of ambipolar diffusion becomes non-negligible can be obtained by solving $\mathcal{R}_{\rm AD} = 1$ that yields

$$\ell_{\rm AD} = 0.09\ {\rm pc}\left(\frac{B}{30\ \mu{\rm G}}\right)^2\left(\frac{n}{10^3\ {\rm cm^{-3}}}\right)^{-3/2}\left(\frac{v}{1\ {\rm km/s}}\right)^{-1}, \tag{4.2}$$

which is comparable to the filament width and suggests that the scale obtained using the ambipolar diffusion affects the filament dynamics. In the context of the solar chromosphere, Snow and Hillier (2021) performed two-dimensional two-fluid simulations of SSI for partially ionized gas. They demonstrated that the neutral fluid stabilizes the SSI on a small scale and found new features such as gas accumulation at valleys. However, the situation in their simulations is different from the one in molecular clouds (e.g., ionization degree, ion-neutral collision cross-section, etc.), and they did not study the dispersion relation and the dependence on density and magnetic field. The linear analysis of SSI including ambipolar diffusion is challenging. Our strategy is to directly simulate the SSI including the ambipolar diffusion and the measurement of the growth rate. As the first step to understanding the effect of SSI on filament width, we study the effect of ambipolar diffusion on the SSI and derive the most unstable scale.

4.2 Setup for Simulations

We perform two-dimensional ideal/non-ideal MHD simulations using Athena++ code (Stone et al. 2020). We use the second-order accurate van Leer predictor-corrector scheme and piecewise linear method applied to primitive variables to integrate the equations. The constrained transport method (Stone and Gardiner 2009) ensures the divergence-free condition, $\nabla \cdot \boldsymbol{B} = 0$. In this paper, we do not solve the Poisson equation for self-gravity because we concentrate ourselves on the physics of SSI under the influence of ambipolar diffusion as a first step of this sort of study. The effect of self-gravity will be taken into account in our future studies. We solve the following equations:

$$\frac{\partial \rho}{\partial t} + \nabla \cdot (\rho \boldsymbol{v}) = 0 \tag{4.3}$$

$$\frac{\partial \rho \boldsymbol{v}}{\partial t} + \nabla \cdot \left(\rho \boldsymbol{v}\boldsymbol{v} - \frac{\boldsymbol{B}\boldsymbol{B}}{4\pi} + P^* \boldsymbol{I} + \boldsymbol{\Pi} \right) = 0 \tag{4.4}$$

$$\frac{\partial E}{\partial t} + \nabla \cdot \left[(E + P^*) \boldsymbol{v} - \boldsymbol{B}(\boldsymbol{B} \cdot \boldsymbol{v}) + \boldsymbol{\Pi} \cdot \boldsymbol{v} + \frac{\eta_{\mathrm{AD}}}{|\boldsymbol{B}|^2} \{ \boldsymbol{B} \times (\boldsymbol{J} \times \boldsymbol{B}) \} \times \boldsymbol{B} \right] = 0 \tag{4.5}$$

$$\frac{\partial \boldsymbol{B}}{\partial t} - \nabla \times \left[(\boldsymbol{v} \times \boldsymbol{B}) - \frac{\eta_{\mathrm{AD}}}{|\boldsymbol{B}|^2} \boldsymbol{B} \times (\boldsymbol{J} \times \boldsymbol{B}) \right] = 0, \tag{4.6}$$

where, $P^* = p + B^2/(8\pi)$ and $E = e + \rho v^2/2 + B^2/(8\pi)$ are the total pressure and the total energy density; ρ, p, $\boldsymbol{v}$, and $\boldsymbol{B}$ are the density, pressure, velocity, and magnetic field; and $\boldsymbol{J} = \nabla \times B$ is the current. We introduce the viscous stress tensor

$$\Pi_{ij} = \rho \nu \left(\frac{\partial v_i}{\partial x_j} + \frac{\partial v_j}{\partial x_j} - \frac{2}{3} \delta_{ij} \nabla \cdot v \right) \tag{4.7}$$

to prevent the carbuncle phenomenon (Quirk 1994; Liou 2000; Kim et al. 2003) and the growth of a grid scale SSI seeded by the carbuncle instability. ν is the coefficient of physical kinematic viscosity. We adjust ν to stabilize a grid scale (eight meshes) fluctuation. The box size L_{box} and ν are chosen so that the stabilizing scale by ambipolar diffusion is sufficiently smaller than this grid scale. η_{AD} is the ambipolar diffusion coefficient, which is given by

$$\eta_{\mathrm{AD}} = \frac{B^2}{4\pi \gamma_{\mathrm{in}} \rho_{\mathrm{n}} \rho_{\mathrm{i}}}, \tag{4.8}$$

where, $\gamma_{\mathrm{in}} \equiv \langle \sigma_{\mathrm{in}} v_{\mathrm{in}} \rangle / (m + m_{\mathrm{i}}) = 3.5 \times 10^{13}\,\mathrm{cm}^3\,\mathrm{g}^{-1}\,\mathrm{s}^{-1}$, $\rho_{\mathrm{n}} \simeq \rho$ and ρ_{i} are neutral gas mass density and ion mass density. σ_{in}, v_{in}, m, and m_{i} are Langevin cross-section, the relative velocity between the neutral molecule and ion, mean molecule mass and mean ion mass, respectively. We can write the ρ_i as $C\rho^{1/2}$ when ionization by cosmic rays balances recombination. In this study we apply $C = 3 \times 10^{-16}\,\mathrm{cm}^{-3/2}\,\mathrm{g}^{1/2}$ (Shu 1992).

We numerically solve Eqs. (4.3)–(4.6) on a two-dimensional domain of the size $[-4L_{\mathrm{box}}, 4L_{\mathrm{box}}] \times [0\,\mathrm{pc}, L_{\mathrm{box}}]$ in the shock rest frame. We choose $L_{\mathrm{box}} = 1$ pc for the $\gamma = 5/3$ case, and $L_{\mathrm{box}} = 0.2, 0.25$, or 0.5 pc for the $\gamma = 1.01$ case. The initial density, velocity, and pressure field are set to be

$$\rho(x, y) = \rho_0 \left[1 + \frac{r-1}{2}(1 - \tanh[x/0.003L_{\mathrm{box}}])\right] + \rho_{\mathrm{p}}(x, y), \tag{4.9}$$

$$v_x(x, y) = -v_{x0} \left[1 + \frac{r-1}{2}(1 - \tanh[x/0.003L_{\mathrm{box}}])\right]^{-1}, \tag{4.10}$$

and

$$p(x, y) = p_0 \left[1 + \frac{r_{\mathrm{pres}}-1}{2}(1 - \tanh(x/0.003L_{\mathrm{box}}))\right] \tag{4.11}$$

respectively, where, ρ_0, v_{x0}, and p_0 are the initial density, x-component of velocity, and pressure in the pre-shock region, respectively. We perform simulations for dozens of sets of ρ_0 and v_{x0}. As a seed of the instability, we set the upstream gas sound speed c_s as $0.2\,\mathrm{km\,s^{-1}}$ so that the p_0 is given by $p_0 = \rho_0 c_s^2/\gamma$. The compression ratio r and the pressure jump r_{pres} are

$$r \equiv \frac{(\gamma+1)\mathcal{M}_s^2}{(\gamma-1)\mathcal{M}_s^2 + 2}, \tag{4.12}$$

and

$$r_{\mathrm{pres}} \equiv \frac{2\gamma\mathcal{M}_s^2 - (\gamma-1)}{\gamma+1}. \tag{4.13}$$

As a seed of instability, we introduce density perturbation as follows.

$$\rho_{\mathrm{p}} = 10^{-4} \times \rho_0 \cos\left(\frac{2\pi y}{\lambda_{\mathrm{p}}}\right) \sin\left(\pi\frac{x - 0.02L_{\mathrm{box}}}{0.01L_{\mathrm{box}}}\right), \text{ if } 0.005L_{\mathrm{box}} \le x \le 0.015L_{\mathrm{box}} \tag{4.14}$$

where, λ_{p} is the wavelength of the perturbation. These initial conditions lead to a perturbed stationary shock at $x = 0$. Since the star-forming filaments are perpendicular to the magnetic field, we set the initial uniform magnetic field along the x-axis $B_0\hat{x}$. The numerical domain is a 2D box with a uniform grid of 4096 × 512 cells, which leads to a spatial resolution of $\Delta x = L_{\mathrm{box}}/512$. We impose free boundary conditions at $x = -4L_{\mathrm{box}}, 4L_{\mathrm{box}}$ boundaries. For the $y = 0, L_{\mathrm{box}}$ boundaries, we use periodic boundary conditions.

We simulate 22 different models. Each model has a unique name, starting with "n" (for "upstream density n_0"), followed by the number density ("800," "1000," "1300," and "1600" [cm^{-3}]), the magnetic field ("b"), followed by the field strength

Table 4.1 Model parameters

Model name	n_0 [cm^{-3}]	B_0 [μG]	v_{x0} [km s^{-1}]	γ	$\mathcal{R}_{\rm shear} \equiv v_{x0}\Delta x/\nu$	Ambipolar diffusion	Solver
n1000b30v1$\mathcal{R}\infty$	1000	30	1.0	5/3	∞	No	Roe
n1000b30v1$\mathcal{R}$16.3	1000	30	1.0	5/3	16.3	No	Roe
n1000b30v1$\mathcal{R}$9.8	1000	30	1.0	5/3	9.8	No	Roe
n1000b30v1$\mathcal{R}\infty$E	1000	30	1.0	5/3	∞	No	HLLE
n1000b30v1$\mathcal{R}\infty$D	1000	30	1.0	5/3	∞	No	HLLD
n1000b30v1$\mathcal{R}$16.3D	1000	30	1.0	5/3	16.3	No	HLLD
n1000b30v1$\mathcal{R}$9.8D	1000	30	1.0	5/3	9.8	No	HLLD
n1000b30v1$\mathcal{R}\infty$LD	1000	30	1.0	5/3	∞	No	LHLLD
n1000b30v1$\mathcal{R}$16.3LD	1000	30	1.0	5/3	16.3	No	LHLLD
n1000b30v1$\mathcal{R}$9.8LD	1000	30	1.0	5/3	9.8	No	LHLLD
n1000b30v1.3a$\mathcal{R}\infty$D	1000	30	1.3	5/3	∞	No	HLLD
n1000b30v1.3a$\mathcal{R}\infty$LD	1000	30	1.3	5/3	∞	No	LHLLD
n1000b30v1	1000	30	1.0	1.01	19.5	No	Roe
n1000b30v1AD	1000	30	1.0	1.01	19.5	Yes	Roe
n800b30v1AD	800	30	1.0	1.01	19.5	Yes	Roe
n1300b30v1AD	1300	30	1.0	1.01	19.5	Yes	Roe
n1600b30v1AD	1600	30	1.0	1.01	19.5	Yes	Roe
n1000b24v1AD	1000	24	1.0	1.01	19.5	Yes	Roe
n1000b35v1AD	1000	35	1.0	1.01	19.5	Yes	Roe
n1000b40v1AD	1000	40	1.0	1.01	19.5	Yes	Roe
n1000b30v0.8AD	1000	30	0.8	1.01	19.5	Yes	Roe
n1000b30v0.9AD	1000	30	0.9	1.01	19.5	Yes	Roe

("24," "30," "35," and "40" [μG]), and the velocity ("v"), followed by the upstream velocity ("0.8," "0.9," '1," and "1.3" [km s^{-1}]), followed by the Reynolds number of the physical shear viscosity ("9.8," "16.3," "19.5," and "∞"). Models with a different MHD solver are additionally denoted as "E," "D," and "LD" corresponding to HLLE, HLLD, and LHLLD, respectively. Models with ambipolar diffusion are additionally denoted as "AD." The set of parameters used in our simulations is listed in Table 4.1.

We also perform simulations for the case with initially perturbed shock position fluctuation, which is convenient for the investigation of nonlinear evolution. The initial density, velocity, and pressure field are set to be

$$\rho(x, y) = \rho_0 \left[1 + \frac{r-1}{2} \left(1 - \tanh \left[\frac{x - \xi \sin(2\pi y/\lambda_{\rm p})}{0.003 L_{\rm box}} \right] \right) \right], \tag{4.15}$$

Table 4.2 Model parameters for the nonlinear evolution of one-shock SSI

Model name	λ_p	Ambipolar diffsion
NL1Shock	0.1	No
NL1ShockAD	0.1	Yes
NL1Shockλ_p0.01AD	0.01	Yes

$$v_x(x, y) = -v_{x0}\left[1 + \frac{r-1}{2}\left(1 - \tanh\left[\frac{x - \xi \sin(2\pi y/\lambda_p)}{0.003L_{\text{box}}}\right]\right)\right]^{-1}, \tag{4.16}$$

and

$$p(x, y) = p_0\left[1 + \frac{r_{\text{pres}}-1}{2}\left(1 - \tanh\left[\frac{x - \xi \sin(2\pi y/\lambda_p)}{0.003L_{\text{box}}}\right]\right)\right], \tag{4.17}$$

where ξ is the initial amplitude of shock position fluctuation. We choose $n_0 = 1000\,\text{cm}^{-3}$, $v_{x0} = 1\,\text{km s}^{-1}$, $p_0 = \gamma\rho_0 c_s^2$, $\mathcal{R}_{\text{shear}} = 9.8$, and $\xi = 0.02$ pc. The initial uniform magnetic field is set to be along the x-axis and the strength of $B_0 = 30\,\mu$G. The set of parameters used in our simulations of the nonlinear evolution of one-shock SSI is listed in Table 4.2.

4.3 Results

4.3.1 Linear Evolution for One-Shock SSI

4.3.1.1 Selection of Numerical Scheme and Physical Viscosity

We perform simulations for the case of $\gamma = 5/3$ and compare the results with dispersion relation (Eq. 1.23), by which we can check whether the selected numerical scheme appropriately reproduces the regime of linear instability. The method to measure the growth rate of SSI is the same as the method developed by Stone and Edelman (1995). It is convenient to use compression weighted averages because we need to evaluate physical quantities in the vicinity of a shock wave. We can write compression-weighted transverse magnetic energy as follows.

$$\langle B_y^2 \rangle = \frac{\int B_y^2 C dV}{\int C dV} \tag{4.18}$$

where,

$$C = \min\left(\partial v_x/\partial x, 0\right). \tag{4.19}$$

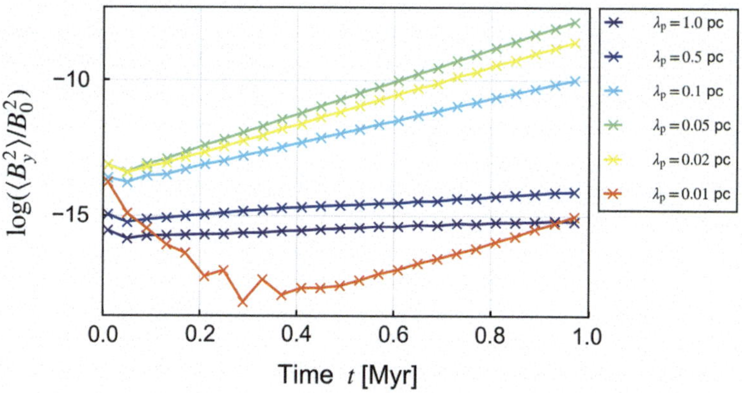

Fig. 4.1 Evolution of the mean value of the perturbed magnetic field for the case with adiabatic ideal MHD including physical shear viscosity (model n1000b30v1$\mathcal{R}$9.8). Reproduced from Abe et al. (2024).

In Fig. 4.1, we show the evolution of $\langle B_y^2 \rangle$ in model n1000b30v1a. The slope of each line in Fig. 4.1 reflects the growth rate. The growth rate can be measured from the slope of $\langle B_y^2 \rangle$ as

$$\omega_{\rm num} = \left(2\log_{10} e\right)^{-1} \frac{d}{dt} \log_{10} \frac{\langle B_y^2 \rangle}{B_0^2} \simeq \left(2\log_{10} e\right)^{-1} \frac{\Delta \log_{10} \langle B_y^2 \rangle / B_0^2}{\Delta t}. \tag{4.20}$$

where, $\frac{\Delta \log_{10} \langle B_y^2 \rangle / B_0^2}{\Delta t}$ is the gradient in the t–$\log_{10} \langle B_y^2 \rangle / B_0^2$ plane. We can confirm linear growth. For $\lambda_{\rm p} = 0.01$ pc, $\langle B_y^2 \rangle / B_0^2$ decreases until ~0.3 Myr, a larger scale noise start to grow after t ~ 0.3 Myr, which is different from the growth of $\lambda_{\rm p} = 0.01$ pc mode. Since the initial perturbation is not given as the eigenstate of the SSI, the SSI does not start to grow $t = 0$. Thus we define the measuring range as $[t_0 = t_{\rm start} + f t_{\rm growth}, t_0 + t_{\rm range}]$ to observe the linear growth of SSI. where, $t_{\rm growth} \equiv 1/\omega_{\rm ana}$ is a growth timescale We choose $t_{\rm ini} = 0.05$ Myr, $t_{\rm range} = 0.2$ Myr and $f = 0.4$.

The HLLD and LHLLD Riemann solvers are robust and high-resolution MHD solvers, but it does not take into account the slow mode characteristics in the Riemann problem, and it is necessary to test the optimal solver for solving SSI. We perform simulations using HLLD, LHLLD, HLLE, and Roe solvers to decide which solver is the best for solving SSI. If physical shear viscosity is not included, the measurement of the growth rate fails due to the carbuncle phenomenon. We can prevent the carbuncle phenomenon by introducing physical shear viscosity. Grid noise, which is numerically induced noise whose scale is different from a given $\lambda_{\rm p}$, potentially becomes the seed of SSI. The growth of SSI seeded by the grid noise may arise after a long time integration (e.g., see the red line in Fig. 4.1), thus we avoid measuring the growth of grid noise by adjusting the $t_{\rm range}$.

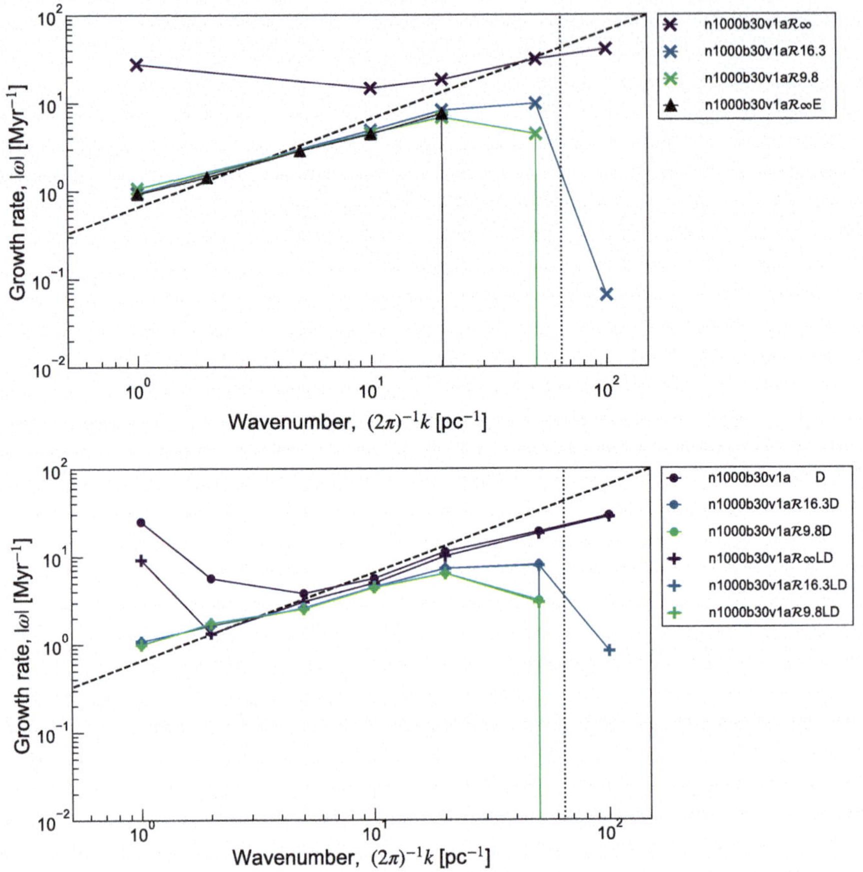

Fig. 4.2 Dispersion relation for adiabatic cases. *Top panel*: Dispersion relations for the case with Roe (cross marker, models n1000b30v1a$\mathcal{R}\infty$, n1000b30v1a$\mathcal{R}$16.3, and n1000b30v1a$\mathcal{R}$9.8) and HLLE (triangle marker, model n1000b30v1a$\mathcal{R}\infty$E) solvers. Black and purple lines show the non-viscous results. Blue and green lines show the result with $\mathcal{R}_{\rm shear} = 16.3$ and $\mathcal{R}_{\rm shear} = 9.8$. *Bottom panel*: the same as top panel, but for the case with HLLD (circle marker, models n1000b30v1a$\mathcal{R}\infty$D, n1000b30v1a$\mathcal{R}$16.3D, and n1000b30v1a$\mathcal{R}$9.8D) and LHLLD (plus marker, models n1000b30v1a$\mathcal{R}\infty$LD, n1000b30v1a$\mathcal{R}$16.3LD, and n1000b30v1a$\mathcal{R}$9.8LD) solvers. Reproduced from Abe et al. (2024).

The top panel of Fig. 4.2 shows the dispersion relation for adiabatic cases calculated using HLLE or Roe solvers (models n1000b30v1a$\mathcal{R}\infty$E, n1000b30v1a$\mathcal{R}\infty$, n1000b30v1a$\mathcal{R}$16.3, and n1000b30v1a$\mathcal{R}$9.8). The dashed line represents Eq. (1.23). The vertical dotted line represents the scale of $\lambda = 8\Delta x$. We can see that the grid scale structure created by the carbuncle phenomenon is suppressed by introducing physical shear viscosity. For simulations employing the HLLE solver, we do not introduce the physical shear viscosity, since the carbuncle phenomenon does not occur, however very diffusive nature of the HLLE solver makes SSI attenuate at $0.02 \leq \lambda_p < 0.05$ pc (see, filled triangles). In the results with the Roe solver without the physical shear viscosity (bold cross marks), we cannot measure growth rates at long wavelength regimes, due to the contamination by the carbuncle phenomenon. For model n1000b30v1$\mathcal{R}$16.3 (blue crosses), the carbuncle phenomenon still appears at $\lambda_p = 0.02$ pc (see the right edge of the blue curve). Using the Roe method with $\mathcal{R}_{\rm shear} = 9.8$ (see green crosses), the SSI growth rate can be measured with high resolution (down to ~0.02 pc) as well as preventing carbuncle phenomenon.

The bottom panel of Fig. 4.2 shows the same as the top but results using HLLD or LHLLD solvers (models n1000b30v1a$\mathcal{R}\infty$D, n1000b30v1a$\mathcal{R}$16.3D, n1000b30v1a$\mathcal{R}$9.8D, n1000b30v1a$\mathcal{R}\infty$LD, n1000b30v1a$\mathcal{R}$16.3LD, and n1000b 30v1a$\mathcal{R}$9.8LD). The simulation using HLLD suffers from the carbuncle phenomenon. The LHLLD scheme is designed to alleviate the carbuncle phenomenon, but after a long time of integration, the growth of grid noise appears. Both schemes give similar results, but we fail to measure for $\lambda_p > 1.0$ pc modes if we do not involve physical shear viscosity by the effects of the carbuncle phenomenon due to slower growth of the SSI than the carbuncle phenomenon (see filled purple circles and purple plus marks). It should be mentioned that the results using LHLLD are closer to the approximated analytical solution than those using HLLD at $\lambda_p \leq 0.5$ pc. Simulations using either HLLD or LHLLD successfully reproduce the growth of SSI for $\mathcal{R}_{\rm shear} = 9.8$.

In a conclusion, we find that the HLLD, LHLLD, and Roe solvers with adjusted physical shear viscosity can correctly calculate the growth rate of SSI over a wide scale range. We can use any of the HLLD, LHLLD, and Roe solvers to measure the linear growth rate, but in the following sections, we use the Roe solver that shows numerically more stable features in nonlinear regimes (see, Appendix).

4.3.1.2 Dispersion Relation of SSI in Molecular Clouds

Since the isothermal treatment is justified in dense regions of molecular clouds, we adopt $\gamma = 1.01$. We measure the growth rate in the same way as Sect. 4.3.1.1. In Fig. 4.3, we show the evolution of the mean value of the perturbed magnetic field for model n1000b30v1. For $\lambda_p = 0.002$ pc, $\langle B_y^2 \rangle / B_0^2$ decreases until ~0.2 Myr, then a larger scale (> 0.002 pc) grid noise grows after $t \sim 0.2$ Myr, which is different from the growth of $\lambda_p = 0.002$ pc mode of the SSI. For $\lambda_p = 0.2, 0.1, 0.02, 0.005$ pc, we can see the slope increments of the perturbed magnetic field after $t = 0.7$ Myr caused by grid noise. $\langle B_y^2 \rangle / B_0^2$ converges to -4 after $t = 0.75$ Myr for $\lambda_p = 0.01$ due to the saturation of SSI (Stone and Edelman 1995). As in Sect. 4.3.1.1, the range to measure the linear growth rate is $[t_0 = t_{\rm start} + f t_{\rm growth}, t_0 + t_{\rm range}]$, and we choose $t_{\rm start} = 0.05$ Myr, $t_{\rm range} = 0.5$ Myr, and $f = 0.6$. We show the dispersion relation for the isothermal ideal MHD case including physical shear viscosity (model n1000b30v1) as the grey cross marks in Fig. 4.5. We find that, if we use $\mathcal{R}_{\rm shear} = 19.5$, we find that a physical dispersion relation is successfully obtained by suppressing the carbuncle phenomenon.

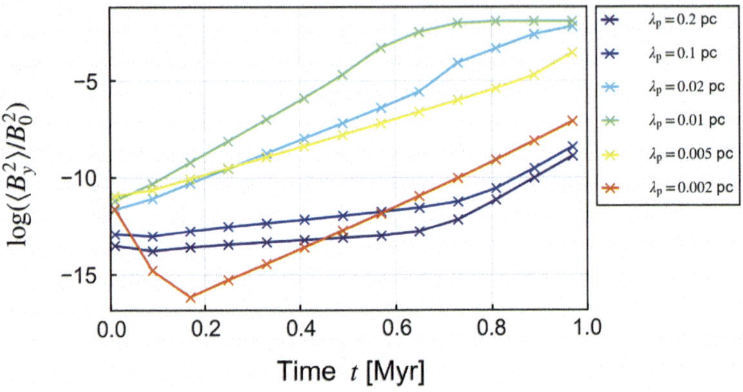

Fig. 4.3 Evolution of the mean value of the perturbed magnetic field for the case with isothermal ideal MHD including physical shear viscosity (model n1000b30v1). Reproduced from Abe et al. (2024).

4.3.1.3 SSI Versus Ambipolar Diffusion

We perform a similar analysis as in Sect. 4.3.1.2 for the simulation results including ambipolar diffusion. In Fig. 4.4, we show the evolution of the mean value of the perturbed magnetic field for model n1000b30v1AD. Because the effect of ambipolar diffusion makes the phase speed of the Alfvén wave smaller, it takes more time for the eigen state to develop from the given initial perturbation compared to the ideal MHD case. Thus, for measuring the liner phase growth rate, we take longer $t_{\rm start}$ of 1.2 Myr, and $f = 0.6$ and $t_{\rm range} = 0.3$ Myr. We show the dispersion relation for the isothermal MHD case including ambipolar diffusion (model n1000b30v1AD) as the black cross marks in Fig. 4.5. We can see the damping of the SSI growth by the ambipolar diffusion. We find that the most unstable scale $\ell_{\rm max} \simeq 0.2$ pc and the damping scale $\ell_{\rm damp} \simeq 0.02$ pc.

To investigate the parameter dependence on the most unstable scale, we perform a parameter survey for unperturbed magnetic field strength, density, and shock velocity. In Fig. 4.6, we show the dispersion relation for models n1000b24v1AD, n1000b40v1AD, n800b30v1AD, n1600b24v1AD, n1000b30v0.8AD. We can confirm that the most unstable scale changes with the magnetic field density and velocity. The larger the density/velocity, the smaller the most unstable scale of ambipolar diffusion. Contrary, the larger the magnetic field, the larger the damping scale. These trends can be understood by the scale of ambipolar diffusion $\ell_{\rm damp} \sim \ell_{\rm AD} \propto B_0^2 \rho^{-3/2} v_{x0}^{-1}$ (Eq. 4.2).

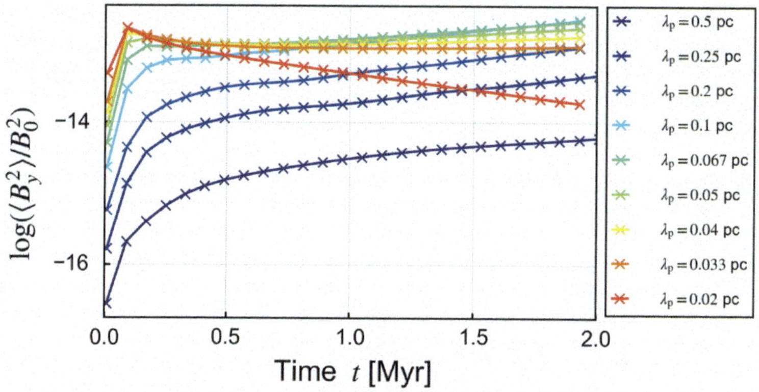

Fig. 4.4 Evolution of the mean value of the perturbed magnetic field for the case with isothermal ideal MHD including ambipolar diffusion and physical shear viscosity (model n1000b30v1AD). Reproduced from Abe et al. (2024). ©2024. The Author(s). Published by the American Astronomical Society. CC BY 4.0

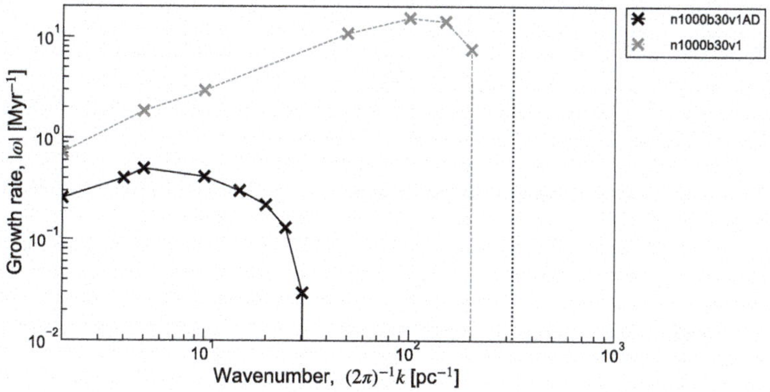

Fig. 4.5 Dispersion relation for the model n1000b30v1 (grey) and n1000b30v1AD (black). Reproduced from Abe et al. (2024). ©2024. The Author(s). Published by the American Astronomical Society. CC BY 4.0

4.3.2 Nonlinear Evolution for One-Shock SSI

We show the density maps of nonlinear evolution for model NL1Shock in Fig. 4.7. The white lines represent the contour of the vector potential, i.e., magnetic field lines. The results are consistent with Stone and Edelman (1995). The preshock gas flows along the magnetic field lines, pushing the fluid away from the peaks and concentrating it in the valleys. This effect reduces/increases preshock densities in peak/valley and provides the observed changes in postshock densities. Changes in preshock density affect shock velocities, causing peak/valley velocities to be higher/lower than average, resulting in finger growth.

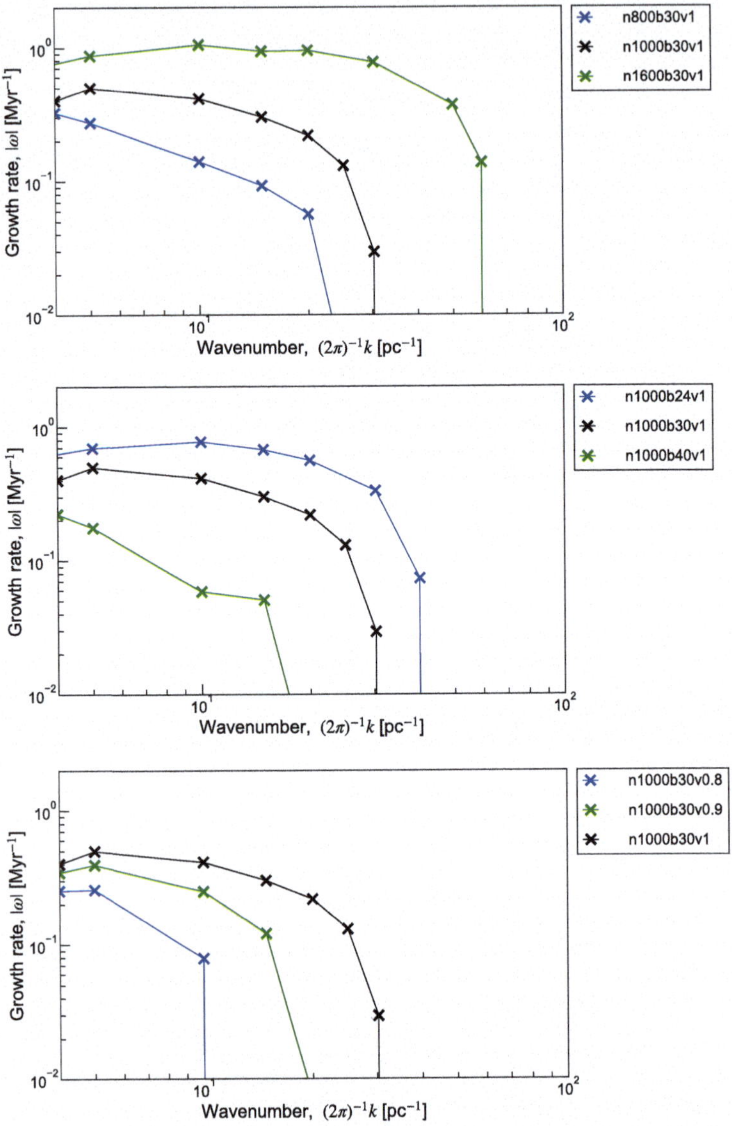

Fig. 4.6 Dispersion relation for the models n1000b30v1AD, n1600b30v1AD, n800b30v1AD, n1000b24v1AD, n1000b40v1AD, and n1000b30v0.8AD. Reproduced from Abe et al. (2024).

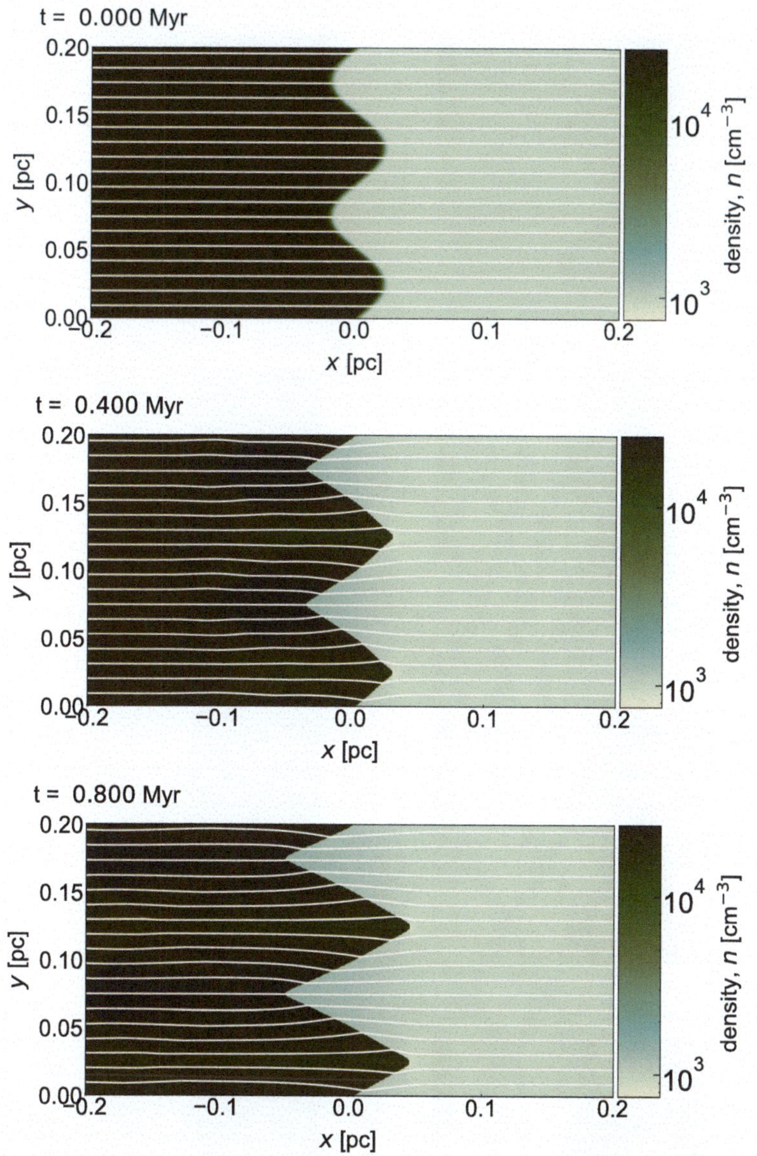

Fig. 4.7 Density maps in the result of model NL1Shock at time t = 0.0, 0.4, and 0.8 Myr (from top to bottom). White lines represent the magnetic field lines. Reproduced from Abe et al. (2024).

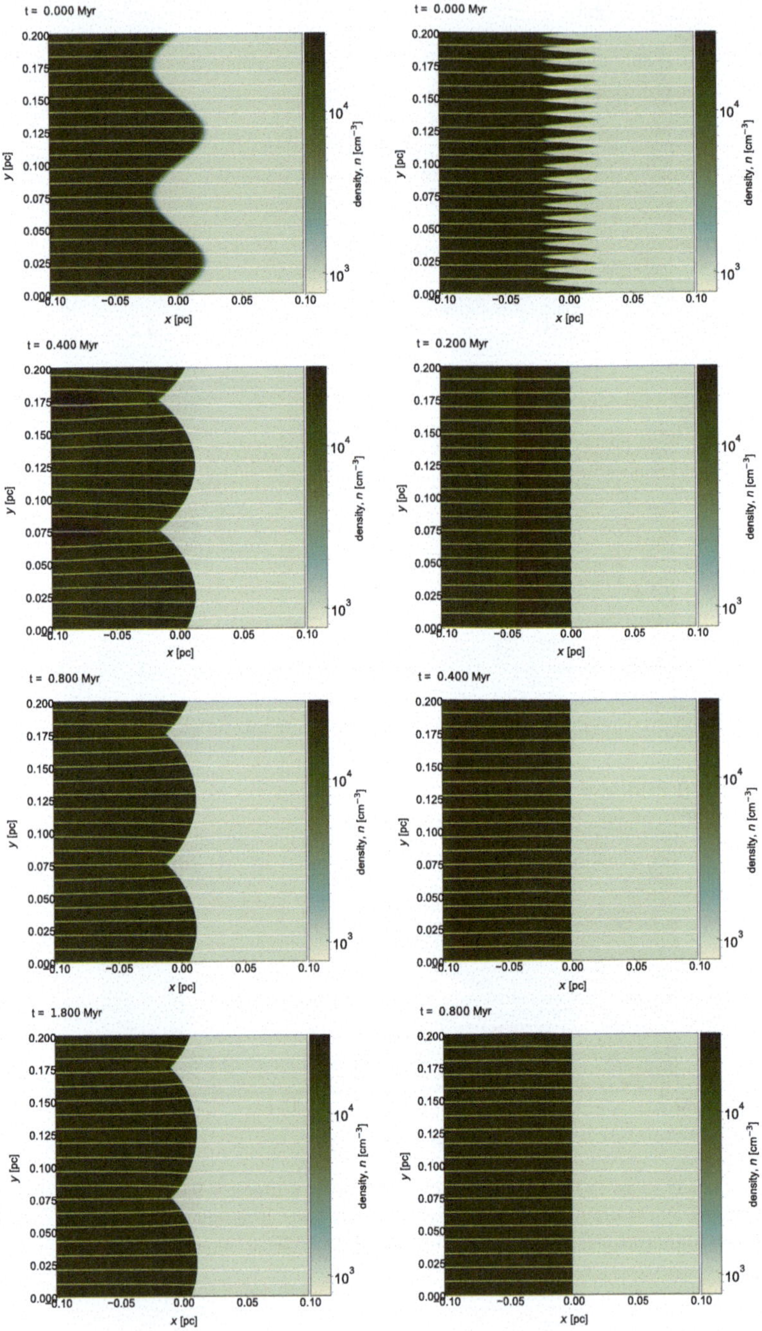

Fig. 4.8 Same as Fig. 4.8 but for model NL1ShockAD (left column) and NL1Shockλ_p0.01AD (right column). Reproduced from Abe et al. (2024).

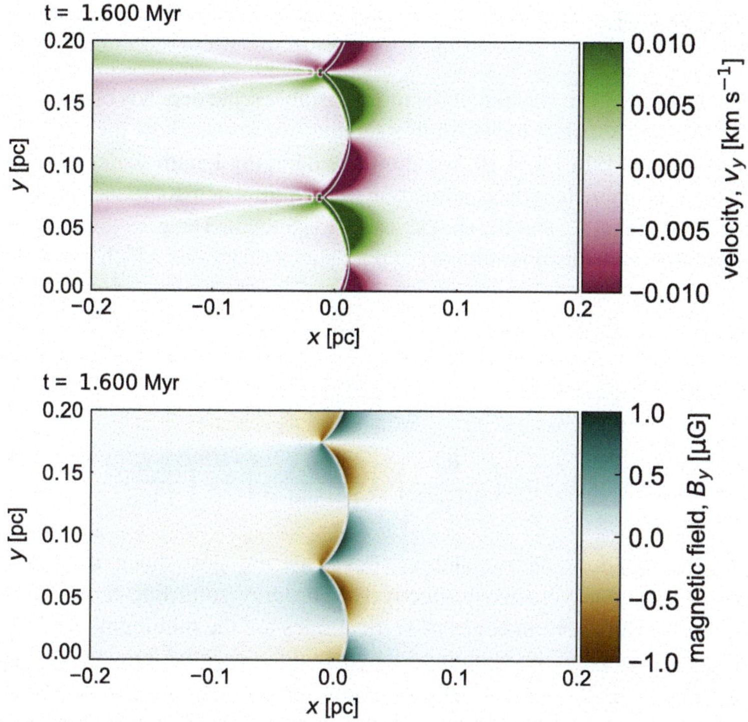

Fig. 4.9 v_y and B_y maps in the result of model NL1ShockAD at time t = 1.6 Myr. Reproduced from Abe et al. (2024). ©2024. The Author(s). Published by the American Astronomical Society. CC BY 4.0

In the left panels of Fig. 4.8, we show the density maps of non-linear evolution including ambipolar diffusion (model NL1ShockAD). We can confirm the growth rate of shock position fluctuation is slowed down due to ambipolar diffusion. The right panels of Fig. 4.8 show the density maps of non-linear evolution including ambipolar diffusion (model NL1Shockλ_p0.01AD). Because $\lambda_p = 0.01$ pc $< \ell_{damp} \simeq 0.03$ pc, i.e., ambipolar diffusion strongly affects, shock position fluctuation is stabilized and disappears. Figure 4.9 shows that the gas flows along the shock front (see top panel), leading to the formation of high-density regions at the valleys (see left panels of Fig. 4.8). The magnetic field structure is close to the eigenstate of slow shock and kinks around the shock front (see a bottom panel or white lines in Fig. 4.8). These indicate that ambipolar diffusion strongly works around the shock front. This result is consistent with Snow and Hillier (2021).

4.4 Discussion

In Sect. 4.3.1.3, we have stated that the most unstable scale depends on the magnetic field, density, and shock velocity. In this section, we discuss how the characteristic of the SSI is scaled. In Fig. 4.10, we show the damping length scale, which scales with the most unstable length, as a function of the initial density (panel a), the initial magnetic field (panel b), and the shock velocity (panel c). The grey dotted line is the scale of ambipolar diffusion evaluated in post-shock quantities which can be written as

$$\ell_{\rm AD,post} \equiv \frac{B_1^2}{4\pi\gamma_{\rm in} C \rho_1^{3/2} v_1} \simeq \frac{B_0^2 c_{\rm s}}{4\pi\gamma_{\rm in} C \rho_0^{3/2} v_0^2} \qquad (4.21)$$
$$= 0.017\ {\rm pc} \left(\frac{B_0}{30\ \mu{\rm G}}\right)^2 \left(\frac{n_0}{10^3\ {\rm cm}^{-3}}\right)^{-3/2} \left(\frac{v_0}{1\ {\rm km/s}}\right)^{-2},$$

where, $B_1 \simeq B_0$, $\rho_1 \simeq \mathcal{M}_{\rm s}^2 \rho_0$, and $v_1 \simeq \mathcal{M}_{\rm s}^{-2} v_0$ are post-shock magnetic field strength, density, and velocity, respectively. The grey solid line represents $1.7 \times \ell_{\rm AD,post}$. The top and bottom edges of vertical lines are the minimum lengths with a positive growth rate and the maximum lengths with a negative growth rate, i.e., the vertical lines show the range in which the damping scale exists. For panels (a) and (b), the damping length scale follows $1.7\ell_{\rm AD,post}$. For the shock velocity dependence, the dependency is not as good as n_0 and B_0, but the difference from $1.7\ell_{\rm AD,post}$ is within a factor of 2. Thus we conclude that the characteristic damping scale of the SSI can be approximately estimated by $\ell_{\rm damp} \sim 1.7\ell_{\rm AD,post}$, and the most unstable scale by $\ell_{\rm SSI} \sim 5\ell_{\rm damp} \sim 9\ell_{\rm AD,post}$.

4.5 Summary

We have performed the SSI simulations in molecular clouds using isothermal ($\gamma = 1.01$) or adiabatic ($\gamma = 5/3$) using Athena++. To test the optimal solver for solving SSI, we performed simulations using HLLD, LHLLD, HLLE, and Roe solvers to decide which solver is the best for solving SSI by comparison with the approximated analytic solution (Eq. 1.23). The stabilization length scale of SSI in molecular clouds by ambipolar diffusion was investigated. We also study the nonlinear evolution of SSI and the effect of ambipolar diffusion on it. Furthermore, we performed a three-dimensional simulation including self-gravity and ambipolar diffusion. Our main findings are as follows.

1. HLLD, LHLLD, and Roe solvers with physical kinetic viscosity are most suitable for simulating SSI. Simulations using HLLD/LHLLD solvers could cause unphysical explosions depending on the initial condition. Although such a numer-

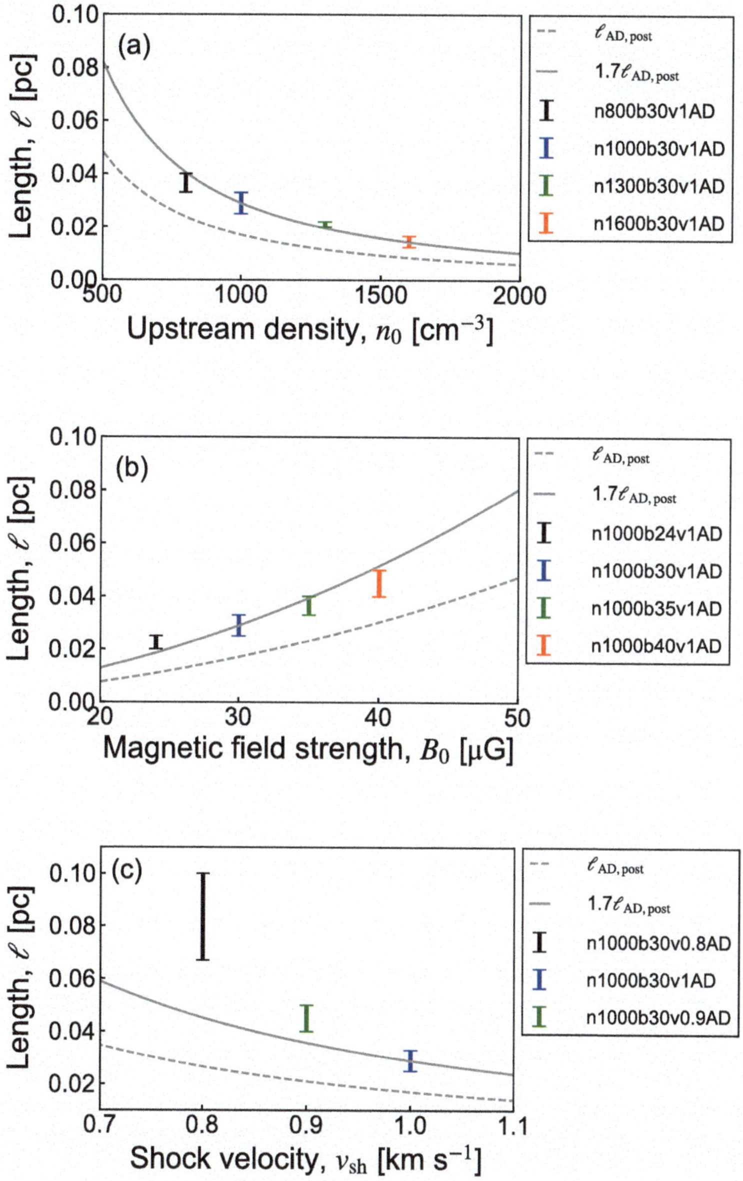

Fig. 4.10 Parameter dependence of damping scale $\ell_{\rm damp}$ (solid vertical lines). The grey dashed and solid lines represent $\ell_{\rm AD,post}$ (see, Eq. 4.21) and $1.7\ell_{\rm AD,post}$. **a** Upstream density versus damping scale. **b** Magnetic field versus damping scale. **c** Shock velocity versus damping scale. Reproduced from Abe et al. (2024).

ical effect can be quenched using physical viscosity, we select the Roe solver as an optimal one in Athena++.
2. Ambipolar diffusion suppresses SSI on a small scale. We find that the most unstable scale of the order of ~0.1 pc and the damping scale of the order of ~0.01 pc in molecular clouds.
3. The most unstable and damping length scales depend on the density, shock velocity, and, magnetic field strength. The scaling is roughly described by $\ell_{\rm SSI} \sim 5\ell_{\rm damp} \sim 9\ell_{\rm AD,post} \propto B_0^2 \rho^{-3/2} v_{x0}^{-2}$.

References

Abe D, Inoue T, Inutsuka S-I (2024) ApJ 961:100. https://doi.org/10.3847/1538-4357/ad072a, https://ui.adsabs.harvard.edu/abs/2024ApJ...961..100A
Arzoumanian D, Shimajiri Y, Inutsuka S, Inoue T, Tachihara K (2018) PASJ 70:96. https://doi.org/10.1093/pasj/psy095, https://ui.adsabs.harvard.edu/abs/2018PASJ...70...96A
Auddy S, Basu S, Kudoh T (2016) ApJ 831:46. https://doi.org/10.3847/0004-637X/831/1/46, https://ui.adsabs.harvard.edu/abs/2016ApJ...831...46A
Bonne L et al (2020) A&A 644:A27. https://doi.org/10.1051/0004-6361/202038281, https://ui.adsabs.harvard.edu/abs/2020A&A...644A..27B
Chen C-Y, Mundy LG, Ostriker EC, Storm S, Dhabal A (2020) MNRAS 494:3675. https://doi.org/10.1093/mnras/staa960, https://ui.adsabs.harvard.edu/abs/2020MNRAS.494.3675C
Federrath C (2016) MNRAS 457:375. https://doi.org/10.1093/mnras/stv2880, https://ui.adsabs.harvard.edu/abs/2016MNRAS.457..375F
Fischera J, Martin PG (2012) A&A 542:A77. https://doi.org/10.1051/0004-6361/201218961, https://ui.adsabs.harvard.edu/abs/2012A&A...542A..77F
Inoue T, Inutsuka S-I (2012b) ApJ 759:35. https://doi.org/10.1088/0004-637X/759/1/35, https://ui.adsabs.harvard.edu/abs/2012ApJ...759...35I
Inoue T, Yamazaki R, Inutsuka S-I, Fukui Y (2012) ApJ 744:71. https://doi.org/10.1088/0004-637X/744/1/71, https://ui.adsabs.harvard.edu/abs/2012ApJ...744...71I
Inutsuka S, Miyama SM (1992a) ApJ 388:392. https://doi.org/10.1086/171162, https://ui.adsabs.harvard.edu/abs/1992ApJ...388..392I
Kim S-S, Kim C, Rho O-H, Kyu Hong S (2003) J Comput Phys 185:342. https://doi.org/10.1016/S0021-9991(02)00037-2, https://ui.adsabs.harvard.edu/abs/2003JCoPh.185..342K
Koch EW, Rosolowsky EW (2015) MNRAS 452:3435. https://doi.org/10.1093/mnras/stv1521, http://adsabs.harvard.edu/abs/2015MNRAS.452.3435K
Liou M-S (2000) J Comput Phys 160:623 https://doi.org/10.1006/jcph.2000.6478, https://ui.adsabs.harvard.edu/abs/2000JCoPh.160..623L
Ntormousi E, Hennebelle P (2019) A&A 625:A82. https://doi.org/10.1051/0004-6361/201834094, https://ui.adsabs.harvard.edu/abs/2019A&A...625A..82N
Orkisz JH et al (2019) A&A 624:A113. https://doi.org/10.1051/0004-6361/201833410, https://ui.adsabs.harvard.edu/abs/2019A&A...624A.113O
Palmeirim P et al (2013) A&A 550:A38. https://doi.org/10.1051/0004-6361/201220500, https://ui.adsabs.harvard.edu/abs/2013A&A...550A..38P
Priestley FD, Whitworth AP (2022) MNRAS 512:1407. https://doi.org/10.1093/mnras/stac627, https://ui.adsabs.harvard.edu/abs/2022MNRAS.512.1407P
Quirk JJ (1994) Int J Numer Methods Fluids 18:155. https://doi.org/10.1002/fld.1650180603, https://ui.adsabs.harvard.edu/abs/1994IJNMF..18..555Q

Shimajiri Y, André P, Ntormousi E, Men'shchikov A, Arzoumanian D, Palmeirim P (2019) A&A 632:A83. https://doi.org/10.1051/0004-6361/201935689, https://ui.adsabs.harvard.edu/abs/2019A&A...632A..83S

Shu FH (1992) The physics of astrophysics, vol II. Gas dynamics

Snow B, Hillier A (2021) MNRAS 506:1334. https://doi.org/10.1093/mnras/stab1672, https://ui.adsabs.harvard.edu/abs/2021MNRAS.506.1334S

Stodólkiewicz JS (1963) Acta Astron 13:30. https://ui.adsabs.harvard.edu/abs/1963AcA....13...30S

Stone JM, Edelman M (1995) ApJ 454:182. https://doi.org/10.1086/176476, https://ui.adsabs.harvard.edu/abs/1995ApJ...454..182S

Stone JM, Gardiner T (2009) New Astron 14:139. https://doi.org/10.1016/j.newast.2008.06.003, https://ui.adsabs.harvard.edu/abs/2009NewA...14..139S

Stone JM, Tomida K, White CJ, Felker KG (2020) ApJS 249:4. https://doi.org/10.3847/1538-4365/ab929b, https://ui.adsabs.harvard.edu/abs/2020ApJS..249....4S

Suri S et al (2019) A&A 623:A142. https://doi.org/10.1051/0004-6361/201834049, https://ui.adsabs.harvard.edu/abs/2019A&A...623A.142S

Tomisaka K (2014) ApJ 785. https://doi.org/10.1088/0004-637X/785/1/24

Chapter 5
Conclusion

Filament formation and evolution in molecular clouds determine the initial condition for star formation. However, there are unsolved problems. We tried to solve the following problems.

1. Theoretical research suggests that shock waves in molecular clouds trigger filament formation. Since several different mechanisms have been proposed for filament formation, the formation mechanism of the observed star-forming filaments requires clarification.
2. Previous studies have demonstrated that strong MHD shock compression allows for the formation of massive cores. However, these studies implicitly assumed a long duration of a shock wave (corresponding to collisions between large clouds). On the other hand, some studies demonstrated that gravitationally bound cores do not form in the case of short duration. Therefore, a systematic study into the effect of shock duration on the resulting filament/core formation is required.
3. Observations show that a massive star cluster formation occurs where the peak of gas column density in a cloud exceeds 10^{23} cm^{-2}. However, the physical origin of the threshold peak column density of 10^{23} cm^{-2} is unknown.
4. Observations show that the filament width takes the universal value of 0.1 pc regardless of their line-mass, even supercritical filaments subject to gravitational collapse. Theoretically, however, the width should shrink for the supercritical filaments.

In Chap. 2, we tackled the problem 1. We performed a series of isothermal MHD simulations of filament formation. We focused on the influences of shock velocity and turbulence on the formation mechanism and identified three different mechanisms for filament formation. The results indicate that when the shock is fast, at shock velocity $v_{sh} = 7$ km s^{-1}, the Type O mechanism substantially works. However, at a slow shock velocity $v_{sh} = 2.5$ km s^{-1}, Type C mechanism becomes important. When both the shock velocities and turbulence are low, Type G becomes important for filament formation. Moreover, we analyzed the line-mass distribution of the filaments and

D. Abe, *Formation and Evolution of Star-Forming Filaments in Molecular Clouds*,
Springer Theses, https://doi.org/10.1007/978-981-97-6414-3_5

showed that strong shock waves can naturally create high-line-mass filaments such as those observed in the massive star-forming regions in a short time. We conclude that the dominant filament formation mode changes with the velocity of the shock wave triggering the filament formation. The filament formation mechanism is understood in a unified way by this work.

In Chap. 3, we investigated the effect of the shock-compressed layer duration on filament/star formation and how the initial conditions of massive star formation are realized, which leads to solving problems 2 and 3. We performed three-dimensional (3D) isothermal MHD simulations with gas inflow duration from the boundaries (i.e., shock wave duration) as a controlling parameter. Filaments formed behind the shock expand after the duration time for short shock duration models, whereas long-duration models lead to star formation by forming massive supercritical filaments. Moreover, when the shock duration is longer than two postshock free-fall times, the peak column density of the compressed layer exceeds 10^{23} cm^{-2}, and the gravitational collapse of the layer causes the number of OB stars expected to be formed in the shock-compressed layer reaches the order of ten (i.e., massive cluster formation). In this sense, a faster shock wave is advantageous because postshock free-fall time becomes smaller when the shock wave is faster.

In Chap. 4, to solve problem 4, focusing on the slow-shock instability (SSI) that occurs at the filament boundaries, we used non-ideal MHD simulations to investigate the mechanism of filament width maintenance. The wavefront of the slow-shock is known to be unstable, and we expect the conversion of the accretion flow ram pressure into thermal/turbulent pressure across the unstable shock front that potentially maintains the filament width. However, since the linear dispersion relation obtained from the ideal MHD analysis shows infinitesimal scale as the most unstable scale. In the scale of dense filaments, the effect of ambipolar diffusion is non-negligible and is expected to suppress the instability at small scales. As a first step, we investigated the SSI's linear dispersion relation, including ambipolar diffusion. We performed two-dimensional MHD simulations to examine the linear stage of the slow-shock instability including the effect of the ambipolar diffusion. We found that the most unstable scale of the SSI is given by roughly 5 times the length scale of the ambipolar diffusion calculated using postshock variables.

Appendix A
Procedure for Derivation of Slow Shock Instability

In this appendix, we show the detailed derivation of the slow shock instability based on Lessen and Deshpande (1967) and Édel'Man (1989). Basic equations of isothermal MHD are given by

$$\rho\frac{\partial \mathbf{U}}{\partial t} + U \cdot \nabla U = -\nabla p - \nabla\left(\frac{B^2}{8\pi}\right) + \frac{1}{4\pi}(\mathbf{B}\cdot\nabla)\mathbf{B}, \tag{A.1}$$

$$\frac{\partial \rho}{\partial t} + \nabla\cdot(\rho\mathbf{v}) = 0, \tag{A.2}$$

$$\frac{\partial \mathbf{B}}{\partial t} = \nabla\times(\mathbf{v}\times\mathbf{B}), \tag{A.3}$$

and

$$p = \rho c_{\mathrm{s}}^2. \tag{A.4}$$

Because we consider a situation in a boundary of star-forming filament, the unperturbed gas flow, and magnetic field are set to be perpendicular to the shock wave (see Fig. A.1). The steady-state jump conditions at $x = 0$ are given by

$$\langle \rho u\rangle = 0, \tag{A.5}$$

$$\langle B_x\rangle = 0, \tag{A.6}$$

and

$$\left\langle \rho u^2 + p\right\rangle = 0, \tag{A.7}$$

where $\langle\cdots\rangle$ and $\hat{n}_k$ denotes the jump in the quantity across the shock and the components of the unit normal $\hat{n}$ along the x_k axis. These equations are the same as unmagnetized fluid. When a shock wave moves with a velocity of W, these jump conditions are rewritten as

D. Abe, *Formation and Evolution of Star-Forming Filaments in Molecular Clouds*, Springer Theses, https://doi.org/10.1007/978-981-97-6414-3

Fig. A.1 Flow configuration

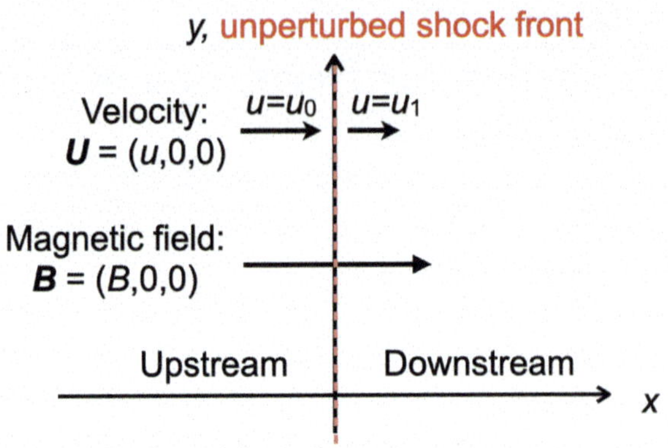

$$- \langle \rho \rangle W + \langle \rho U_i \rangle \hat{n}_i = 0 \tag{A.8}$$

$$\langle B_i \rangle \hat{n}_i = 0 \tag{A.9}$$

$$- \langle \rho U_i \rangle W + \left\langle \rho U_k U_i \hat{n}_k \right\rangle = - \left\langle p + \frac{B^2}{8\pi} \right\rangle \hat{n}_i + \frac{1}{4\pi} \langle B_i B_k \rangle \hat{n}_k \tag{A.10}$$

$$- \langle B_i \rangle W + \langle B_i U_k - B_k U_i \rangle \hat{n}_k = 0 \tag{A.11}$$

We set perturbations as follows.

$$\begin{aligned}
\delta u &= f \exp\left[i \left(k_x x + k_y y - \omega t\right)\right] \\
\delta v &= \Phi \exp\left[i \left(k_x x + k_y y - \omega t\right)\right] \\
\delta B_x &= b_x \exp\left[i \left(k_x x + k_y y - \omega t\right)\right] \\
\delta B_y &= b_y \exp\left[i \left(k_x x + k_y y - \omega t\right)\right] \\
\delta \rho &= n \exp\left[i \left(k_x x + k_y y - \omega t\right)\right] \\
\delta p &= c_s^2 \delta \rho,
\end{aligned} \tag{A.12}$$

where k_x, k_y, and ω are x and y components of wave number and the frequency of disturbance, respectively. The linear equations can be written as

$$\omega_L n + \rho \left(k_x f + k_y \Phi\right) = 0 \tag{A.13}$$

$$\omega_L \rho f + k_x \pi = 0 \tag{A.14}$$

$$\omega_L \rho \Phi + k_y \pi + (B/4\pi) \left(k_y b_x - k_x b_y\right) = 0 \tag{A.15}$$

$$\omega_L b_x + k_y B \Phi = 0 \tag{A.16}$$

$$\omega_{\rm L} b_y - k_x B \Phi = 0 \tag{A.17}$$

$$k_x b_x + k_y b_y = 0 \tag{A.18}$$

where, $\omega_{\rm L} \equiv -\omega + k_x u$. To find a non-trivial solution to these equations, taking the determinant of coefficients to be zero, we get the dispersion relation

$$\omega_{\rm L}^4 - \omega_{\rm L}^2 \left(k_x^2 + k_y^2\right)\left(c_{\rm s}^2 + v_{\rm A}^2\right) + \left(k_x^2 + k_y^2\right) k_x^2 v_{\rm A}^2 c_{\rm s}^2 = 0, \tag{A.19}$$

where, $v_{\rm A} \equiv B/\sqrt{4\pi\rho}$. While the component of the wave number vector along the shock wave, k_y, is the same on both upstream and downstream, the perpendicular component k_x is different on different sides. These values are obtained from A.19. The eigenvalue obtained from the second factor of Eq. A.19 corresponds to magneto-acoustic waves. The shock wave is said to be unstable if there exists a disturbance vanishing at $x = \pm\infty$ and such that Im $\omega > 0$. It is assumed that k_y is real. The boundary condition requires that Im $k_x < 0$ ahead of the shock and Im $k_x > 0$ behind the shock. Since the solutions of the Eq. A.19 are complex, we solve it by applying approximation of $\mathcal{M}_{\rm A} \ll 1$ and $\omega/(k_y u) \ll 1$ Édel'Man (1989). Multiplying $\mathcal{M}_{\rm A}^2 (k_y u)^4/\omega^4$ Eq. A.19 by, we obtain

$$(-\Omega + l)^4 - (-\Omega + l)^2 \left(l^2 + l\right)\left(M_s^{-2} + M_A^2\right) + \left(l^2 + 1\right) l^2 \mathcal{M}_A^{-2} \mathcal{M}_s^{-2} = 0 \tag{A.20}$$

where, $\Omega \equiv \omega/(k_y u)$ and $l \equiv k_x/k_y$. An upstream solution that does not diverge at infinity is $k_x = -ik_y$. Solutions in downstream are $k_x = ik_y$ and $k_x = \Omega k_y/(1 + \mathcal{M}_1)$.

The eigen functions for a magneto-acoustic wave are given by

$$f = -\frac{c_{\rm s}^2 k_x}{\omega_{\rm L}} \frac{n}{\rho} \tag{A.21}$$

$$\Phi = -\frac{c_{\rm s}^2 k_y \omega_{\rm L}^2}{\omega_{\rm L}\left[\omega_{\rm L}^2 - \left(k_x^2 + k_y^2\right) v_{\rm A}^2\right]} \frac{n}{\rho} \tag{A.22}$$

$$b_x = \frac{c_{\rm s}^2 k_y^2 B_x}{\omega_{\rm L}^2 - \left(k_x^2 + k_y^2\right) v_{\rm A}^2} \frac{n}{\rho} \tag{A.23}$$

$$b_y = -\frac{c_{\rm s}^2 k_y k_x B_x}{\omega_{\rm L}^2 - \left(k_x^2 + k_y^2\right) v_{\rm A}^2} \frac{n}{\rho} \tag{A.24}$$

The surface of the perturbed shock is written as

$$\xi = \eta \exp\left(ik_y y - i\omega t\right). \tag{A.25}$$

This leads to the shock velocity normal to itself,

$$W = -i\omega\xi. \tag{A.26}$$

Note that the shock amplitude η is small. The unit normal vector to the perturbed surface is given by $\hat{n}_k = \left(1, -ik_y\xi\right)$. Using these equations and the shock jump condition, we can obtain the jump condition for the perturbation

$$i\left\langle\rho\omega - k_y\rho v\right\rangle\xi + \langle\rho f + un\rangle = 0 \tag{A.27}$$

$$\langle b_x\rangle = 0 \tag{A.28}$$

$$\left\langle 2\rho u f + u^2 n + \pi\right\rangle = 0 \tag{A.29}$$

$$i\left\langle -k_y\rho\left(v^2 + c_s^2\right)\right\rangle\xi + \left\langle \rho u\Phi - \frac{B_x b_y}{4\pi}\right\rangle = 0 \tag{A.30}$$

$$\left\langle -B_x\Phi + ub_y\right\rangle = 0 \tag{A.31}$$

Substituting the eigen functions to Eqs. A.28 and A.31, we can confirm that Eqs. A.28 and A.31 are identical, i.e., there are four independent equations.

A solution of the density perturbation can be written as a superposition of each eigen function. For the slow shock, the solutions upstream and downstream are given by

$$n_0 = n_0^{(\mathrm{f})} \exp\left(ik_{x_{\mathrm{fast}0}} x\right), \tag{A.32}$$

and

$$n_1 = n_1^{(\mathrm{f})} \exp\left(ik_{x_{\mathrm{fast}1}} x\right) + n_1^{(\mathrm{s})} \exp\left(ik_{x_{\mathrm{slow}1}} x\right). \tag{A.33}$$

Substituting the eigen functions and Eqs. A.32 and A.33 to the perturbed shock jump condition, the perturbed jump condition can be written as the following form

$$\| S_{ij} \| \begin{vmatrix} n_1^{(f)} \\ n_2^{(f)} \\ n_2^{(s)} \\ \eta \end{vmatrix} = 0 \tag{A.34}$$

where $\| S_{ij} \|$ is a 4×4 matrix. The determinant of S_{ij} gives a non-trivial solution, that is dispersion relation of slow shock instability.

Appendix B
High Shock Velocity Simulations

B.1 High Shock Velocity Case with Strong Turbulence

In the model v12GyTy, the velocity dispersion of the turbulence is set to 1 km s^{-1}. In this appendix, we investigate the filament formation mechanism and the distribution of the angle between the filament and the magnetic field lines in the case of larger velocity dispersion of the initial turbulence $\Delta v = 2$ km s^{-1} as model 'v12GyTy'. Panels (I)–(III) and (i)–(iii) in Fig. B.1 show column density snapshots of model v12GyTy2 in the y-z and x-y planes at $t = 0.2$, 0.4, and 0.8 Myr, respectively. We can confirm that stronger turbulence makes the structure more complex than the results in model v12GyTy. To clarify the dominant filament formation mechanism, we show the local density cross-sections around the five major filaments as the results of models v12GyTy2 in Fig. B.2. The snapshots show that turbulent motion parallel to the magnetic field (y-direction) create dense clamps in the pre-shock region[1] that are swept by the shock and induces type O filament formations. The curved shock morphology and velocity vectors (black arrows) shown in the cross-section panels in model v12GyTy2 support the activation of the type O mechanism.

We show angle histograms of model v12GyTy2 at $t = 0.2$ (panels a and b), 0.4 (panels c and d), and 0.8 (panels e and f) Myr in Fig. B.3. Left panels (a), (c), and (e) are the angle histograms for the filaments in the column density range of 0.5 $\bar{N}_{\rm sh}$ to 1.5 $\bar{N}_{\rm sh}$. Right panels (b), (d), and (f) are the histograms for the filaments with $N > 1.5\ \bar{N}_{\rm sh}$. While dense filaments are perpendicular to the magnetic field, the distribution of the angle between the faint filament and the magnetic field is random and different from that in Fig. 2.14. Remarkably, the histograms of the angle distribution are much flatter than the case of model v12GyTy (see, Fig. B.3). This may be because, when the initial turbulence is strong, various filamentation modes are mixed and the resulting filamentary structure changes its direction to the magnetic

[1] This resembles type C filament formation, but this process happens before shocked sheet formation, and thus preshock clumps are formed.

D. Abe, *Formation and Evolution of Star-Forming Filaments in Molecular Clouds*, Springer Theses, https://doi.org/10.1007/978-981-97-6414-3

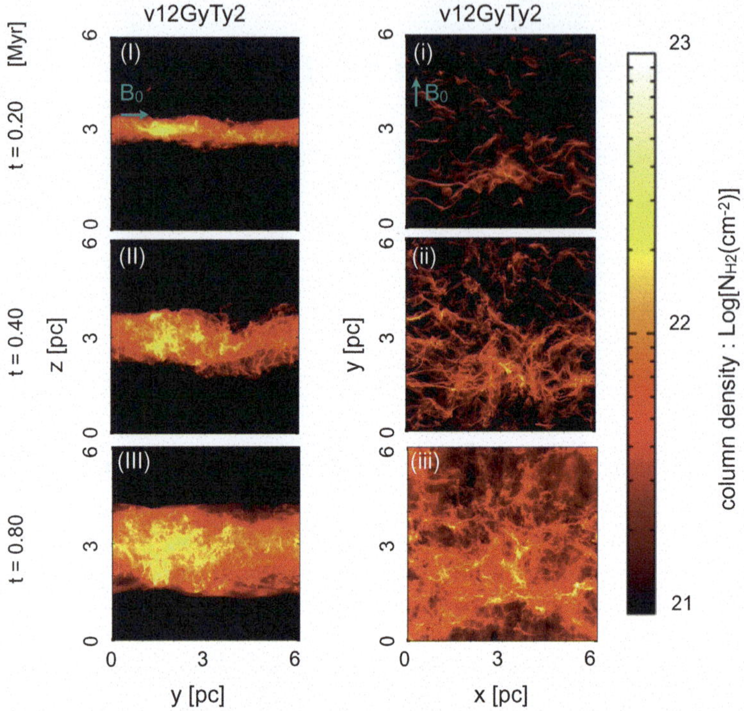

Fig. B.1 Column density maps at time $t = 0.2$ (*top*), 0.4 (*middle*), and 0.8 (*bottom*) Myr. *Left row* (panels I, II, and III): Column density in the *y*-*z* plane of model v12GyTy2. *Right row* (panels i, ii, and iii): Same as the panels (I)–(III) but for the *x*-*y* plane. Reproduced from Abe et al. (2021).

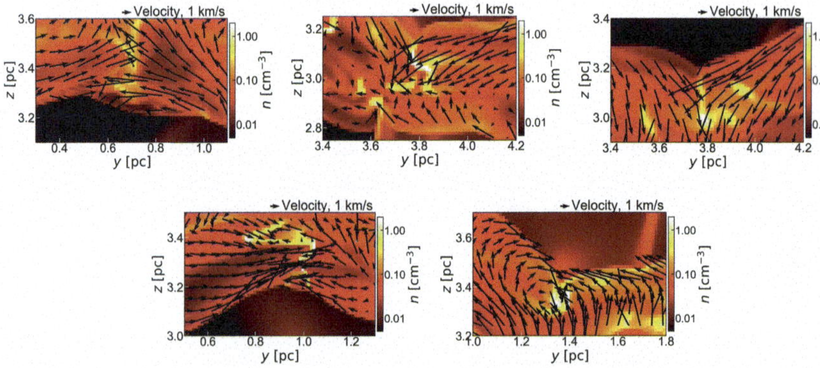

Fig. B.2 Cross-section maps of the number density in the *y*-*z* plane of model v12GyTy2. The yellow blobs located roughly at the center of each panel correspond to cross-sections of the filaments. We can confirm that the oblique MHD shock compression mechanism takes a major role in filament formation even in the case of initial stronger turbulence with high shock velocity. Reproduced from Abe et al. (2021).

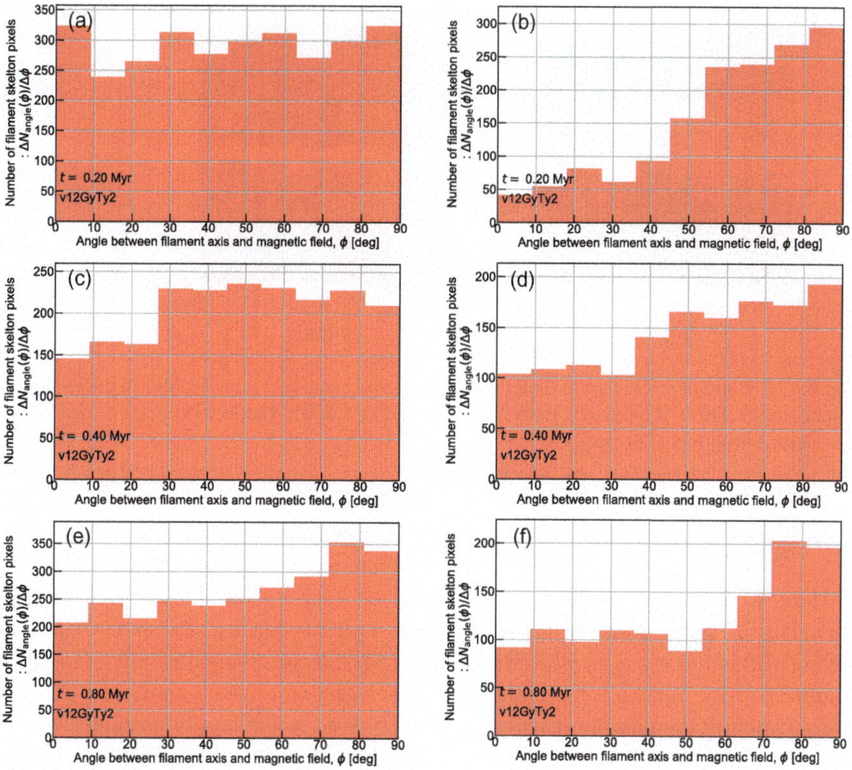

Fig. B.3 Histogram of angles between filaments and magnetic field for model v12GyTy2. From top to bottom, results at time $t = 0.2, 0.4$, and 0.8 Myr, respectively. *Left panels* (1), (3), and (5): Results when we identify filaments in the column density range of $0.5\bar{N}_{\rm sh}$ to $1.5\bar{N}_{\rm sh}$. *Right panels* (2), (4), and (6): Results when the filament identification threshold column density is chosen to be $1.5\bar{N}_{\rm sh}$. Reproduced from Abe et al. (2021). ©2021. The Author(s). Published by the American Astronomical Society. CC BY 4.0

field lines. Observational study of density structure and magnetic field orientations in several molecular clouds by Planck Collaboration XXXV (2016) does not show such a flat distribution. Therefore, we conclude that this kind of initial strong turbulence model does not provide a realistic model.

B.2 High Shock Velocity Case without Magnetic Field

In this section, we describe the additional set of calculations without a magnetic field. The compression ratio of the isothermal hydrodynamics shock is given by $R \sim \mathcal{M}_{\rm s}^2$, where $\mathcal{M}_{\rm s}$ is the sonic Mach number. Thus, for instance, if we set $v_{\rm sh} = 7.0$ (or 2.5) km s^{-1}, the compression ratio becomes 1225 (or 156), which is extremely high.

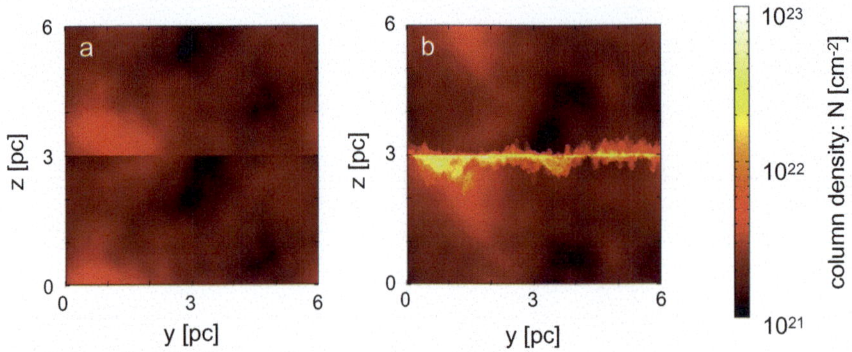

Fig. B.4 Column density maps at time $t = 0.3$ $(panela)$, and 1.0 $(panelb)$ Myr in the y-z plane of model v8GnTnB0. Reproduced from Abe et al. (2021). ©2021. The Author(s). Published by the American Astronomical Society. CC BY 4.0

Note that, with the magnetic field, the compression ratio is dramatically reduced to $R \sim \sqrt{2}\mathcal{M}_A$, where $\mathcal{M}_A$ is the Alfvénic Mach number.

Here, we chose parameter $v_{coll} \sim 8.0\,\mathrm{km\ s^{-1}}$ ($v_{sh} \sim 5\,\mathrm{km\ s^{-1}}$), which we call v8GnTnB0. Panels a and b in Fig. B.4 show column density snapshots of model v8GnTnB0 in the y-z at $t = 0.3$ Myr, and 1.0 Myr, respectively. In contrast to other column density figures in the main text, we integrate the whole numerical domain in constructing the column density. In panel b, we see the highly unstable feature of shock surfaces. This seems to be nonlinear thin-shell instability (Vishniac 1994), which is known to be suppressed in the case of a magnetic field. A similar feature is observed in previous similar studies (e.g., Folini et al. 2014). In both panels, compressed layers sandwiched by two shocks are extremely thin and dense, which has not been observationally identified in the actual molecular clouds associated with fast shock waves as far as we are aware. Under this high compression, the length scale of the initial fluctuations becomes smaller than the spatial resolution, and thus, the smallest scale structures shown in this section are limited by numerical resolution. Nonetheless, we can identify no sign of the formation of the filamentary structure with a sufficiently large line-mass. Therefore, we think that this setup without a magnetic field is not realistic and the presence of magnetic fields is critical for creating filamentary structures in the shock-compressed region.

Appendix C
Selection of MHD Solver

In Figs. C.1 and C.2, we show snapshots of the density and pressure map of models n1000b30v1.3a$\mathcal{R}\infty$D and n1000b30v1.3a$\mathcal{R}\infty$LD, respectively. For long-term simulations using HLLD/LHLLD without the physical shear viscosity, numerical errors around the shock front cause unphysical numerical explosions, which do not appear for simulations with the Roe scheme. This numerical problem occurs if the denominator $\rho_\alpha\,(S_\alpha - u_\alpha)\,(S_\alpha - S_M) - B_x^2$ in Eq. (44)–(47) of Miyoshi and Kusano (2005) closes to zero. The latest Athena++ has been designed to prevent this issue to some extent, but it cannot prevent the unphysical explosion under the initial conditions dealt with in this study. Although such a numerical effect can be quenched by physical viscosity, we chose the Roe solver with physical shear viscosity for the safety of long-term integration.

D. Abe, *Formation and Evolution of Star-Forming Filaments in Molecular Clouds*, Springer Theses, https://doi.org/10.1007/978-981-97-6414-3

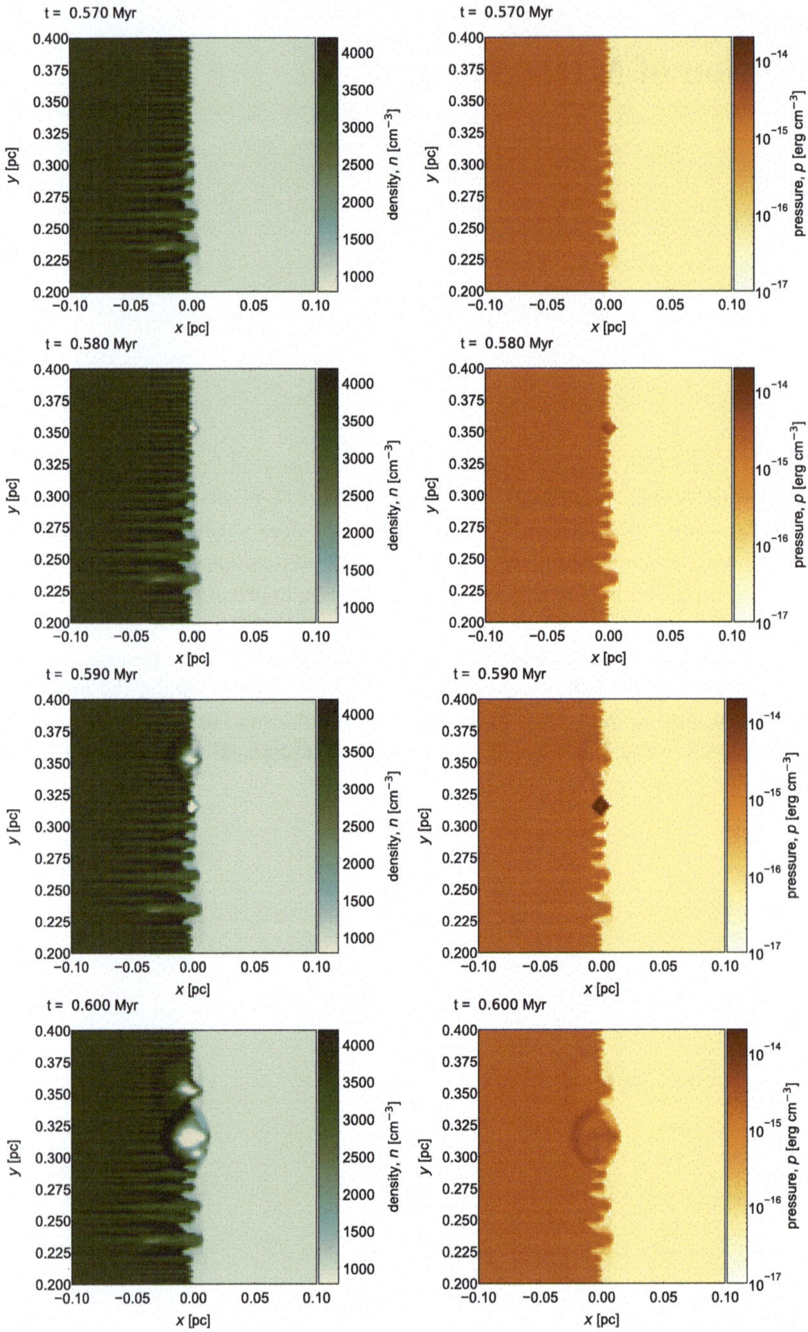

Fig. C.1 Density (left row) and pressure (right row) maps in the result of model n1000b30v1.3a at time $t = 0.57$, 0.58, 0.59, and 0.60 Myr (from top to bottom). Reproduced from Abe et al. (2024).

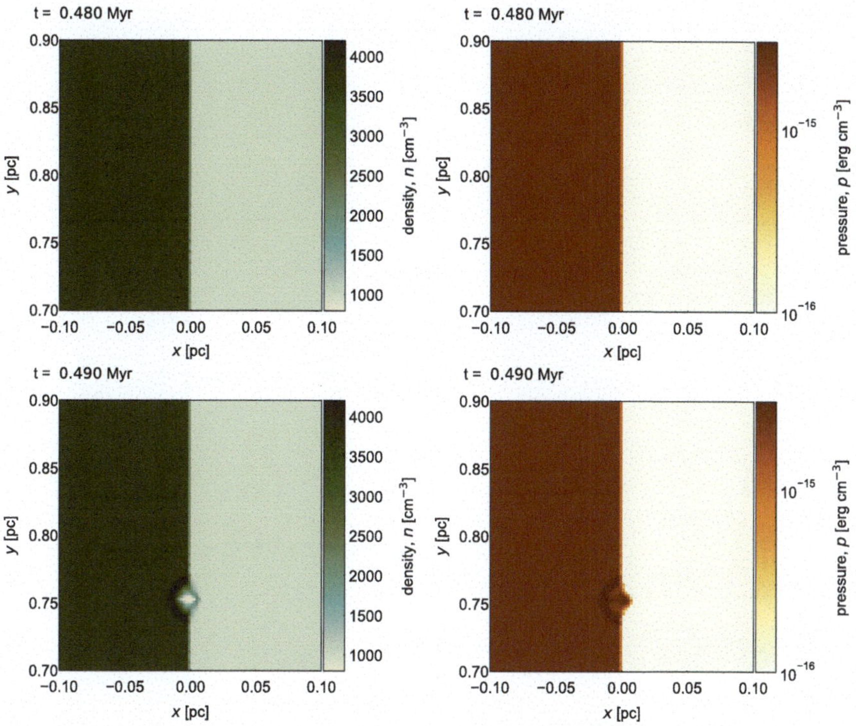

Fig. C.2 Density (left row) and pressure (right row) maps in the result of model n1000b30v1.3a at time $t = 0.57$, 0.58, 0.59, and 0.60 Myr (from top to bottom). Reproduced from Abe et al. (2024). ©2024. The Author(s). Published by the American Astronomical Society. CC BY 4.0

References

Abe D, Inoue T, Inutsuka S-I, Matsumoto T (2021) ApJ 916:83
Abe D, Inoue T, Inutsuka S-I (2024) ApJ 961:100
Édel'Man MA (1989) Astrophysics 31:758
Folini D, Walder R, Favre JM (2014) A&A 562:A112
Lessen M, Deshpande NV (1967) J Plasma Phys 1:463
Miyoshi T, Kusano K (2005) [JCoPh] 208:315. https://doi.org/10.1016/j.jcp.2005.02.017
Planck Collaboration XXXV (2016) A&A 586:A138
Vishniac ET (1994) ApJ 428:186

The manufacturer's authorised representative in the EU is Springer Nature Customer Service Centre GmbH, Europaplatz 3, 69115 Heidelberg, Germany. If you have any concerns regarding our products, please contact ProductSafety@springernature.com

Printed and bound by CPI Group (UK) Ltd, Croydon, CR0 4YY
01/10/2026
02225980-0004